Powering the Eagle...Over 90 Years and Counting

Pratt & Whitney's Inspirational Women

Powering the Eagle... Over 90 Years and Counting

Pratt & Whitney's Inspirational Women

American Institute of Aeronautics and Astronautics
12700 Sunrise Valley Drive, Suite 200
Reston, VA 20191-5807

1 2 3 4 5

ISBN: 978-1-62410-383-4

Dedication

This book is dedicated to all working women of Pratt & Whitney, past and present. It is inscribed to all of you because you have been the foundation for the book's origins and growth. It was with all the admiration and affection that has been distilled from your timeless efforts and continuous contribution, that a small team of dedicated women were inspired to tell your stories.

Whether resurrected from disintegrating pieces of paper or copied from a digital archive, the stories will not be forgotten.

These pages are for you.

Acknowledgments

We would like to acknowledge the tremendous support of Pratt & Whitney Vice President and Executive Sponsor for the Women's Council, Mary Anne Cannon, without whom this book would not have been possible.

Additional Thanks

The Book Production Team would also like to thank the following contributors, each providing a necessary contribution to ensure our publishing success:

Nicole E. Gathy, Pratt & Whitney Assistant Intellectual Property Counsel—for your outstanding publishing contract development support and for voluntarily providing our content review to ensure Pratt & Whitney's intellectual property was protected.

Jody Valet, American Rosie the Riveter Association, Treasurer—for reaching out to the association to help us find some of the Pratt & Whitney "Rosies," providing real context through interviews about the World War II era.

Professional Assistance

Special thanks to our publishing partner, American Institute of Aeronautics and Astronautics, especially David Arthur, Acquisitions & Development Editor; Pat DuMoulin, Senior Editor of Books; Toni Ackley, Project Manager; and Karl Ackley, Writer, whose assistance and guidance brought our vision to reality.

The authors and contributors offer our apologies in advance for any omission in this work. With so many dedicated women working at Pratt & Whitney over the decades it is impossible to capture all of their amazing stories.

Lastly, we would like to thank the thousands of Pratt & Whitney women who give flight to the world every day and support a bright future for aerospace.

Foreword

From the beginning, aviation has been a field open to all. For more than 90 years, the women of Pratt & Whitney have been an integral part of that diversity, shaping and forming our industry-leading engines as well as the company's culture.

We capture history to share stories, remember events and document learnings. It is part of our human nature to look at where we've been, where we are now and where we want to go. This book is a celebration of individual contributions over the years and a unique tribute to the many hard-working women, many pictured here, who have been involved in aerospace development. Included are many interesting historical details, personal stories and an abundance of scrapbooked photographs, illustrations and interviews that demonstrate the role of women in the evolution of our proud legacy of engine development, manufacturing and maintenance.

Although the emphasis of this work is on the ever-changing professional roles of women at Pratt & Whitney, I am certain it will hold broad appeal to anyone fascinated by the world of aerospace technology, as well as the cultural development of women's exploration in the related Science, Technology, Engineering, and Math (STEM) fields.

From the beginning of my journey at Pratt & Whitney, I experienced an open field for myself and fellow women interested in aviation and aerospace manufacturing. My career is a testament to the fact that women at Pratt & Whitney have every opportunity to reach their personal level of achievement and play a critical role in the top ranks of the company.

I am honored to have been asked to share my thoughts alongside women who have made such significant contributions. Just as I am in awe of Pratt & Whitney's game-changing technologies, I am equally and continually astounded by today's workforce of women innovators, "product builders" and pioneering leaders. My hope is that this book will inspire countless young girls and women to strive for and achieve their own mark in the world of aviation and aerospace manufacturing, leading us to the next level of exploration in powered flight.

Mary Anne Cannon
Pratt & Whitney Vice President and Executive Sponsor for the Women's Council

Contents

To provide an accurate depiction of the changing cultural perception of women in the workplace, a number of articles included in this book were copied exactly as they appeared in their archival source. Each is denoted with this Pratt & Whitney logo. To some, these original articles may seem inappropriate or outdated by today's organizational standards. They serve as a testament to the cultural trends of the day. As this book progresses, the reader will be able to see how different perceptions of women workers were and how far women have come in their professional journey.

Preface

Powering the Eagle began when the idea to tell Pratt & Whitney's history, from the women's perspective, percolated at the close of a routine work meeting. By early 2015, the idea had sprung into a plan of action with a small team of highly enthusiastic volunteers. By late spring, the book project became real. Mary Anne Cannon, as the Executive Chair for Pratt & Whitney's Women's Council (one of the company's eight Employee Resource Groups), provided a small budget, and Hollie Barnas, as the Women's Council Vice President, procured and contracted a publisher, thereby officially launching the project in September 2015.

Throughout the fall, Priscilla Ubysz (manager of Pratt & Whitney Library Services), Holly Fuzzard (Pratt & Whitney librarian) and Hana Sognnaes (archivist) swept the Pratt & Whitney archives for bound volumes of company happenings. Together, hours were spent finding the stories that colorfully described an earlier era. As the collected paper piles grew, a new member, Janina Rivera (Pratt & Whitney Media Services), joined the production team to support the future information transformation – from paper to pixels. In parallel, the recent era of women's formal technology and innovation recognition data collection was supported by two additional team members. Through their leadership and involvement in STEM promotion, Lauren Brumbaugh and Arianna Barrenechea (Pratt & Whitney engineers and Women's Council members) provided the last elements of the women of Pratt & Whitney's storyline. It is telling that a few inspired women took it upon themselves to pursue this labor of love, empowered by Pratt & Whitney's history of allowing its employees to pursue their goals.

The organization of *Powering the Eagle* allows the reader to easily follow the evolution of Pratt & Whitney's engine development timeline while observing the changing role for women in the aerospace workforce. The book is divided into four chapters. Opening with Pratt & Whitney Aircraft Company's founding in the early 1920s, the book progresses through early engine firsts, the World War II era, peacetime engine developments and, finally, modern aerospace of the 21st century. The stories are told as they were written, politically correct for each included time period, serving as reminders of the professional tone of the times.

A timeline of Pratt &Whitney milestones for the past 90-plus years opens Chapters 1 and 3. Each chapter contains an assortment of stories that either describe a historic event, provide historical evidence to support women's roles in aviation and aerospace or relate biographical information about individuals who were propitious in shaping women's professional history. Overall, a detailed story of the advancement of science and technological careers is shown in a readable and entertaining style. Each chapter is filled with pictures and colorful quotes from people of that era. First-hand accounts through interviews provide deeper insight into what, in some history books, is just a listing of factual information. As evidenced by the stories recounted throughout this volume, women's role in aviation began as pilots. During wartime, women were part mechanic, part inventor and part adventurer in order to thrive in the aerospace workplace. A number of firsts and changing roles are evident in the photos—especially as equality in technical knowledge advanced in post-wartime America. The storyline changed again as women became formally recognized for their technological and manufacturing achievements in the aerospace industry.

The evolution and participation of the working woman throughout Pratt & Whitney's history was certainly not static. Transitioning ably from office help to temporary wartime heroes to, eventually, permanent professionals, the women of *Powering the Eagle* have endeavored to make their indelible mark on the aerospace industry.

Production Team Volunteers

Hollie Barnas: Initiator and Publication Lead

I love stories. I also love taking ideas and creating something from them that hasn't been created before. When the idea to tell the 90-year story of women's contributions and their changing roles at Pratt & Whitney crystalized into the concept of a book, a passion was ignited. And serendipitously, my involvement in Pratt & Whitney's Women's Council, STEM promotion, women's retention project and mentoring created the perfect mix and timing to transform the idea into a reality. I am delighted to have become a storyteller for the women of Pratt & Whitney, especially before the early history is lost or the present gets buried in electronic files. I hope we have done these women proud and that this work will give rise to a vision of opportunity for the next generation of "Pratt women."

Holly Fuzzard: Librarian

Within the overall history of aviation—and the microcosm of Pratt & Whitney—there are many stories to be told of how women have contributed to science and technology in particular and the workforce in general. It is critical that we bring these stories to light in order to illustrate just how far we've come in a few short decades—the perceptual evolution from "girls" featured in company newsletter pieces oriented toward their appearance, hobbies or support roles to a shift in paradigm toward recognizing women who are achieving roles in leadership and innovation is quite interesting.

Janina Rivera: Media and Content Design Lead

One of the thrilling parts of my job at Pratt & Whitney is having access to our vast library chronicling the company's 90-plus years, where the company's proud history comes to life before your eyes through photographs, documents and graphics. Our expansive archives bring the past into clear view, illustrating so many of Pratt & Whitney's proud moments throughout the decades. One particularly pleasing discovery I made concerned the important role women have played in the company's success. I have been honored to take part in this legacy project, and feel that the story will now be told so all can appreciate it.

Hana Sognnaes: Archivist

My pleasure in this project stems from my lifelong love of history and my strong mother who began her professional life in 1930 in Bergen, Norway. She continued in her job throughout the Nazi occupation of Norway during World War II, and I learned from her how women have always played an unheralded role in the defense of freedom. Though the iconography of that war is usually only of "Rosie the Riveter," I knew there had to be a richer and more varied story beyond those posters. When presented with the opportunity to organize the archives, I was astonished by the outstanding depth of Pratt & Whitney's history as documented by the company publications, *The Bee Hive* and *The Powerplant*. With pages and pages of photos and articles devoted to the women who were contributing to the development of our dependable engines, these company publications are treasure troves showing the role played by women over the years. No one had ever looked at their content from the perspective of how the Pratt & Whitney women brought value to the company. I am honored to be able to be part of the recognition of women and their contribution to the continued success of the company, and it is a pleasure to share my joy and pride in what they have accomplished.

Priscilla Ubysz: Researcher and Library Services Lead

Several years ago, I saw posters of Pratt & Whitney women at work from the early years to modern times. The women in the posters reflected the progression of their movement from the factory floor to the office, advancing through professional pathways and, finally, into leadership roles. They reminded me of my mother who, although not at Pratt & Whitney, began working at 16 in a similar factory, rising up proudly to an office position halfway through her 48 years at that company. She was an early role model for working women, and I saw the women at Pratt & Whitney through the same eyes, leading the way for all of us to follow. As I became entrenched in the archives and read through our history, I noticed that the stories changed from talking about the "girls" to how Pratt & Whitney women impacted the aerospace industry itself. I realized that all of the company's women's voices should be heard, loud and clear. As we move swiftly, confidently and strongly toward the future, we should, and must, show how these extraordinary women were critical in making Pratt & Whitney the leader in aerospace that it is today.

Lauren Brumbaugh and Ariana Barrenechea: Women of Innovation, STEP Award content contributors

Lauren: I volunteered because I wanted to help tell the stories of the women who work at Pratt & Whitney. All of their achievements, past and present, should be recognized as a shaping hand for generations to come.

Ariana: My reason for participating on this team is to celebrate those women who paved the way for us to have the opportunity to be successful in this industry. The company has become significantly more inclusive since its founding, and it is in no small part due to the women that came before us. I hope to commemorate the legacy of the women of Pratt & Whitney.

Pratt and Whitney

1860-1910

1860: Francis A. Pratt and Amos Whitney found a new manufactured precision machine tool company, with headquarters in Hartford, Connecticut.

1861–1865: The Pratt & Whitney Machine Tool Company fabricates tools for the gun-making machinery in use by the Union Army during the American Civil War.

1920s

1925: Frederick Rentschler approaches the Pratt & Whitney Tool Company for funding and use of the Pratt & Whitney name for a new venture.

1925 (August): Frederick Rentschler and his small team set up Pratt & Whitney Aircraft in Hartford, Connecticut, in part of the Pratt & Whitney Machine Tool complex on Capitol Avenue.

1925 (December): First run of the R-1340 Wasp.

1926 (May): First flight of the Wasp.

1926 (June): First run of the R-1690 Hornet.

1927 (May): First flight of the Hornet.

1928, 1929 (January): Wasp- and Hornet-powered aircraft perform flawlessly on the first major deployment of the USS Saratoga and the USS Lexington.

1929: William Boeing and Chance Vought lead formation of United Aircraft and Transport Corporation.

1929 (February): Pratt & Whitney Canada begins operations with 10 employees.

1929 (November): First run of the R-985 Wasp Junior.

1930s

1930: Pratt & Whitney completes move begun in 1929 to its new factory in East Hartford, Connecticut.

1930: Wasp Junior–powered Laird wins Thompson trophy.

1931: Gee Bee Model Z with a Wasp Junior wins the Thompson.

1931 (April): First run of the R-1830 Twin Wasp.

1932 (July): First run of the R-2060 Yellow Jacket, Pratt & Whitney's first liquid-cooled design. The engine never flew.

1934: "Air Mail Scandal" leads to the breakup of United Aircraft and Transport. Pratt becomes part of the new United Aircraft Corporation that included Hamilton Standard and Chance Vought-Sikorsky.

1937 (September): First run of the R-2800 Double Wasp.

1939 (July): First flight of the Double Wasp.

Family Tree

1940: Orders from France and England along with President Roosevelt's call for 50,000 airplanes a year lead to a massive expansion of Pratt & Whitney. By 1943, the company would have 9 million square feet of space and employ 40,000 people, compared to 3,000 in 1938.

1940: Pratt & Whitney begins its first turbine engine research with the PT1. Work ended when the government required Pratt & Whitney to concentrate on piston engine development and production.

1940 (November): General Hap Arnold allows Pratt & Whitney to end all liquid-cooled engine development. The H-3130 had only ground tests beginning in April 1940 and never flew.

1941 (April): First run of the R-4360 Wasp Major.

1942 (July): Groundbreaking for massive Kansas City plant to build Double Wasps.

1945: Pratt & Whitney workforce declines to about 26,000.

1945 (August): V-J Day. Pratt & Whitney backlog plummets from $400 million to $3 million. Pratt & Whitney and licensees had built 363,619 engines to support the war effort.

1945 (October): American Airlines orders 252 new Twin Wasps for DC-4s, the biggest commercial order for Pratt & Whitney to date.

1947 (July): Design work begins on the J42 Turbo Wasp after an agreement with Rolls-Royce to license Nene engine technology. First run would be in March 1948 and Navy qualification in October 1948.

1948: Design work begins on the more powerful J48 based on the Rolls-Royce Tay. First flight would be in fall 1949.

1949 (February 26–March 2): B-50 Lucky Lady first round-the-world flight using aerial refueling.

1949 (May): J57 design begins.

1950 (January): First run of the J57.

1951 (August): Pratt & Whitney initiates work on a nuclear aircraft engine.

1952 (April): First flight of the B-52 prototype. The engine won Luke Hobbs and Pratt & Whitney the 1952 Collier Trophy.

1953: First launch of the Rocketdyne-powered Redstone.

1953 (May): The Pratt-powered North American F100 Super Saber becomes the first production aircraft to exceed Mach 1 in level flight.

1955 (spring): First flight of the JT4/J75 engine.

1956 (July): Experimental JT10 first run, Pratt & Whitney's first turbofan design.

1957 (fall): First run of the hydrogen-powered 304 engine.

1957 (December): First launch of the Rocketdyne-powered Atlas.

1958 (February): JT4D becomes Pratt & Whitney's first turbofan engine.

1959 (May): Florida Research and Development Center formally opens.

1940s

1950s

World War II Licensee Emblems

(continued p 82)

THE
PRATT & WHITNEY
AIRCRAFT
COMPANY.
GENERAL OFFICE

CHAPTER ONE

THE EARLY YEARS

Pratt & Whitney engines proved crucial in the careers of women aviators throughout the 1920s and 1930s. The likes of Jacqueline Cochrane, Amelia Earhart, Laura Ingalls, Mae Haizlip, Louise Thaden and Blanche Noyes all benefited from the engines that Pratt & Whitney produced.

The year 1925 proved to be a watershed year for aviation history. Seeking to leverage his experience as a former executive at Wright Aeronautical, Frederick Rentschler, the founder and president of Pratt & Whitney Aircraft, rallied a core group of brilliant specialists to turn spare machine tool factory space into what would become Pratt & Whitney Aircraft.

With Rentschler's foresight, the engineering prowess of George Mead, and the dedicated service from staff like Helen Shockley and Mary Conner, Pratt & Whitney Aircraft enjoyed early success during the inter-war period of aviation. In 1925 the U.S. Navy awarded a contract to Pratt & Whitney, based on the submitted prototypes' stellar power-to-weight ratio and outstanding reliability. More than 11,000 engines, known as the Wasp, were produced between 1926 and 1936. Pratt & Whitney and the Wasp engine benefited from the emergence of women in the office. However, it would be another decade before the factory floor would see "ladies" on the assembly line.

Shockley and Conner were not the only women who contributed to the early years of Pratt & Whitney's accomplishments. Other women were also vital to the company's initial success. Unfortunately, the personnel archives from this era are sparse and do not fully document women's roles during the 1920s and 1930s.

To provide an accurate depiction of the changing cultural perception of women in the workplace, a number of articles included in this book were copied exactly as they appeared in their archival source. Each is denoted with this Pratt & Whitney logo. To some, these original articles may seem inappropriate or outdated by today's organizational standards. They serve as a testament to the cultural trends of the day. As this book progresses, the reader will be able to see how different perceptions of women workers were and how far women have come in their professional journey.

Pratt & Whitney Pioneers

A small group of visionary men and women, among whom was Rentschler, the chairman of the United Aircraft Corporation, learned the lesson of our aeronautical unpreparedness for World War I. These pioneers of the aircraft industry remembered 1917 and devoted themselves to a solution. Our nation was ill prepared for the conflict, least of all in the field of aerospace.

In the spring of 1925, the idea that was to become Pratt & Whitney Aircraft was conceived, and a few months later, Rentschler approached the Department of the Navy regarding a possible engine contract. The Navy suggested if a 400-horsepower engine could be demonstrated on an endurance test, more engines would be ordered and tested. If in-flight demonstrations proved successful, quantity orders might follow.

Rear (left to right): C. W. Deeds, G. J. Mead, F. B. Rentschler, Harry Gunberg, J. J. Borrup; Front (left to right): A. V. D. Willgoos, E. A. Ryder, Lawrence Castonguay, R. M. Campbell, Daniel Jack, W. J. Levack,

REAR (LEFT TO RIGHT): DAVID ROBINSON, HENRY CUDWORTH, FRANK IRMISCHER, WILLIAM P. JONES, ANNA MAY REARDON, MARIE WILKINSON; FRONT (LEFT TO RIGHT): CHARLES PETERSON, ALEX HAKANSON, D. L. BROWN, NORRIS KING, PHILIP TREFFERT

Thus, in August 1925, Pratt & Whitney was organized and began operation in an idle building that had been used temporarily as a storehouse for tobacco. Within a few days, Pratt & Whitney Aircraft could boast of 25 employees, including Anna May Reardon and Marie Wilkinson.

The archival information on women is limited; this photo provides the sole photographic evidence of the women who were involved in the very early days of the company. By late December, the first Pratt & Whitney Wasp engine of 410 horsepower was finished and running. In March 1926 the Wasp passed the rigid Navy endurance tests.

1925

Achieving the World's Speed Record

Men did not have a monopoly on the record books through the 1930s. Women also achieved their own aviation milestones. Amelia Earhart is shown here standing beside her Pratt & Whitney-powered Lockheed aircraft in which she broke the world's speed record for women in 1932.

AMELIA EARHART AND LOCKHEED AIRCRAFT

Flying a Pratt & Whitney Wasp-powered airplane from Los Angeles to Newark in 17 hours, 7 minutes, and 30 seconds, Earhart broke her own transcontinental record by almost two hours.

In general, the early connection between Pratt & Whitney and women was with pioneering women aviators, not within the factory.

1932–33

AMELIA EARHART

Winning for Women

Diminutive in the cockpit of the Wasp Junior–powered Wedell-Williams aircraft used in the Thompson race by Lee Gelbach, Mrs. Mae Haizlip did a good job in winning the Shell Speed Dash for women at a 168.216 mile-an-hour clip. Her speed this year was nearly 85 miles an hour slower than the record she made last year in Cleveland, a discrepancy best explained by the lack of suitable equipment in Los Angeles for women to fly.

1933

Mae Haizlip

Earhart Has Some Competition: Laura Ingalls

LAURA INGALLS GREETS WELL-WISHERS AT FLOYD BENNETT AIRPORT

On September 12, 1935, Laura Ingalls set a new transcontinental non-stop flight record for the fair sex when she raced from Los Angeles to Floyd Bennett Airport, New York, in 13 hours, 34 minutes, and 5 seconds. Her time was more than three and a half hours better than that of the former record holder, Amelia Earhart. She now holds both the eastward and westward transcontinental records. On all her record breaking journeys, she has used a Lockheed monoplane, equipped with a Pratt & Whitney Wasp engine and a Hamilton-Standard propeller.

1935

Women at the Air Races

LEFT TO RIGHT: LOUISE THADEN, BLANCHE NOYES, AND VINCENT BENDIX

Mrs. Louise Thaden and Mrs. Blanche Noyes won first place in the Bendix Race. It was the first time in the history of this event that a woman or, more accurately, two women were victorious. Flying together in a standard Whirlwind-powered Beechcraft cabin biplane, their time from New York to Mines Field (called Los Angeles Municipal Airport for purposes of the races) was 14 hours, 54 minutes, and 49 seconds.

A second team of airwomen, Amelia Earhart and Helen Richey, came in fifth with Miss Earhart's new Wasp-powered Lockheed Electra "flying laboratory" in a total elapsed time of 16 hours, 35 minutes, and 30 seconds.

1936

The Remarkable Laura Ingalls

Laura Ingalls, who learned to fly after Lindbergh crossed the Atlantic, continued to prove herself as one of the most capable women pilots. Among her list of accomplishments was an aerial jaunt to South America that established several records. She flew alone, making the 17,000-mile tour that lasted eight weeks and covered 23 countries. It was the longest solo trip ever made by a woman pilot, the first time a woman had piloted a plane around South America, and the first time a woman had flown over the forbidding Andes Mountains.

Ingalls finished second in a Wasp-powered Lockheed Orion monoplane not quite an hour behind the first-ever women winners of the Bendix Race, Louise Thaden and Blanche Noyes.

1936

Laura Ingalls

The Wasp engine used by Laura Ingalls and other women racers

NR

Smile of Success: Bendix Trophy Win

Jacqueline Cochran's AP-7A was a specially built racer, modified from the original AP-7 powered by a 1,829.39-cubic-inch-displacement (29.97-liter) air-cooled, supercharged Pratt & Whitney Twin Wasp S1B3-G (R-1830-11) two-row 14-cylinder radial engine, with a takeoff power rating of 1,000 horsepower at 2,600 rpm, and normal power rating of 850 horsepower at 2,450 rpm and 5,000 feet (1,524 meters). It turned a three-bladed Hamilton Standard controllable-pitch propeller through a 3:2 gear reduction. The engine had a dry weight of 1,320 pounds (595 kilograms). This is the same airplane in which Cochran won the 1938 Bendix Trophy Race.

Jacqueline Cochran

Cochran Conquers Women's Speed Record

Jacqueline Cochran's Republic fighter

Jacqueline Cochran realized a five-year ambition on September 21, 1937, when she established an International Women's Speed Record of 292 miles per hour at Detroit, Michigan. She flew a Twin Wasp–powered Seversky Executive airplane equipped with a Hamilton Standard constant speed propeller. In 1940, Cochran flew her Twin Wasp–powered Republic fighter at a record speed of 330 miles per hour over a 2,000-kilometer (1,242-mile) course.

The humble beginnings of a hardscrabble, poverty stricken life in Florida did nothing to temper Cochran's enthusiasm for flying. Toiling as a youngster in a beauty parlor, the vibrant flame of her aviation dreams never dimmed. Through perseverance and not taking no for an answer, she became a pilot in 1932.

Through the balance of the 1930s, Cochran would claim many awards including those crowning her as the "Outstanding Woman Flyer in the World," and, in 1938, "The Person Making the Greatest Contribution to Aviation" that year. However, this decade would mark just the start of her amazing career as a pilot, patriot and pillar of women's aviation.

1937

United Aircraft Girls' Club

A United Aircraft Girls' Club was organized recently, and out of the 195 charter members, six comely officers were chosen. Membership in the club has been restricted to girls in the East Hartford Divisions of United Aircraft.

The enthusiasm for the aviation industry buoyed the women and men of Pratt & Whitney, indeed all of America, through the disorienting times of the Great Depression. Aviation offered a promise of what (with enough imagination, zeal and good old American know-how) was possible for those of even the most meager of means.

If the Great Depression proved a daunting test of Pratt & Whitney's durability through difficult times, the women of the Eagle would again be called upon to summon all of their resolve in facing that crucible of horrors...World War II.

1939

Left to right: Mary O'Leary, Second Vice-President; Bertha McNeill, Recording Secretary; Florence Thompson, Treasurer; Mary Moran, Corresponding Secretary; Marion Dexter, First Vice-President; and Ellenore Manke, President

Girls' Club Expands to Kansas City

With the United States fully committed to winning World War II, a new United Aircraft plant was created in Kansas City, Missouri, to help meet ever-increasing production demands.

The new facility became the production site for Pratt & Whitney's premier R-2800 Double Wasp engine, which was introduced in 1939 with first flight occurring the following year.

With the war raging overseas, the Kansas City plant did its part in promoting the iconic image of women in the wartime workforce by creating a new girls' club membership, collectively coined as Rosie the Riveter.

D-431

CHAPTER TWO

THE WAR YEARS

The onset of World War II required profound changes in traditional gender roles. The Japanese bombing of Pearl Harbor on December 7, 1941, and the subsequent declaration of war put into motion the greatest employment initiative since the formation of the Works Progress Administration (WPA) in 1935 by President Roosevelt. With more than 10 million men being drafted during the war years, America saw its military industrial output severely hampered by the reallocation of much of its workforce into a fighting force. The woman factory worker was imagined and swiftly integrated into the United States' war-winning strategy.

Just as Jacqueline Cochrane, considered the first female pilot in the U.S. Air Force, broke multiple barriers in Pratt & Whitney–powered planes, women displayed the same tenacity and grit in the workplace. The flexibility and initiative of the woman employee was evident in every facet of factory life. They were welders, riveters, office machine repairers, chauffeurs, and counselors, just to name a few of the special roles they filled. No matter how eccentric the idea seemed of a woman filling the most specialized of positions, it was done. The nation owes a debt of gratitude to all those women who answered the call with an enthusiasm that served our country well throughout the war years.

As shown in this chapter, the "girls" of Pratt & Whitney in the factory were characterized as equal parts pin-up and production personnel.

New Jobs for Women

The natural feminine attributes of patience, nimbleness, and dexterity have made their female services invaluable in lacing, stitching and upholstering operations on wing panels, fuselages and other fabric-covered parts of an airplane. Additionally, nearly 600 women are operating drill presses, compressed air drills, riveting machines and spot welding machines, as well as performing scores of burring and filing operations necessary to the forming of small parts. Present indications show that the efficiency of women in their new roles is quite equal to that of men whom they are now replacing. The women approach their work with an intelligent seriousness.

1942

Neatly uniformed messenger girls are responsible for the delivery of mail as well as the escorting of visitors, pose in front of an F4U-1 fighter.

New jobs, proficiently performed by women among other things, the operation of the air rivet squeeze.

Northwest's Traffic Agents

Northwest Airlines is introducing the feminine element to a line of work heretofore noted for its male personnel as rugged as the terrain over which it flies. Two-score attractive and accomplished young women have been trained and will take up their duties as flying hostesses on new Pratt & Whitney–powered airplanes. At the same time, the company has announced the appointment of three young women as traffic agents in its Chicago office and plans to follow the same practice in all its offices throughout the country, enabling women to have an exciting career in aviation that previously was the sole domain of men.

Northwest Airlines' fleet consists of Lockheed Electras fitted with Wasp Junior engines and Hamilton Standard propellers. One of the oldest air transport systems in the country, Northwest has been in continuous operation since 1927 and has flown more than 50 million passenger miles without a passenger fatality, a feat for which it was recently publicly honored by the National Safety Council. Such an enviable safety record engenders a trust from the flying public and allows for greater investment by all airlines in Pratt & Whitney products.

1942

Women Begin to Play an Important Part in Production

When World War II interrupted the normal routine of American life by calling many men into the armed services—and demanding tremendous war production as well—the aviation industry transitioned from an almost entirely male factory staff to one in which women were occupying increasingly important and extensive roles.

Women were engaged in widely varied factory activities— inspection, machine-tool operation, and laboratory work. Women first gain experience and become familiar with the factory routine. Later, as they demonstrate ability and proficiency, they are instructed in the more complex duties of handling various production machines. Woman also were busy in the inspection and the analysis of materials.

The average woman factory worker at Pratt & Whitney was typical of the many thousands of her kind in war industries of the time. She may come from the city or from the farm. She may have had a college degree or a high school education. Whatever her background, age or appearance, her mind was fixed on one purpose: to do her part in the war effort.

1942

This woman is turning the flange on bronze bushings.

By working on burring benches, women quickly become acquainted with various engine parts.

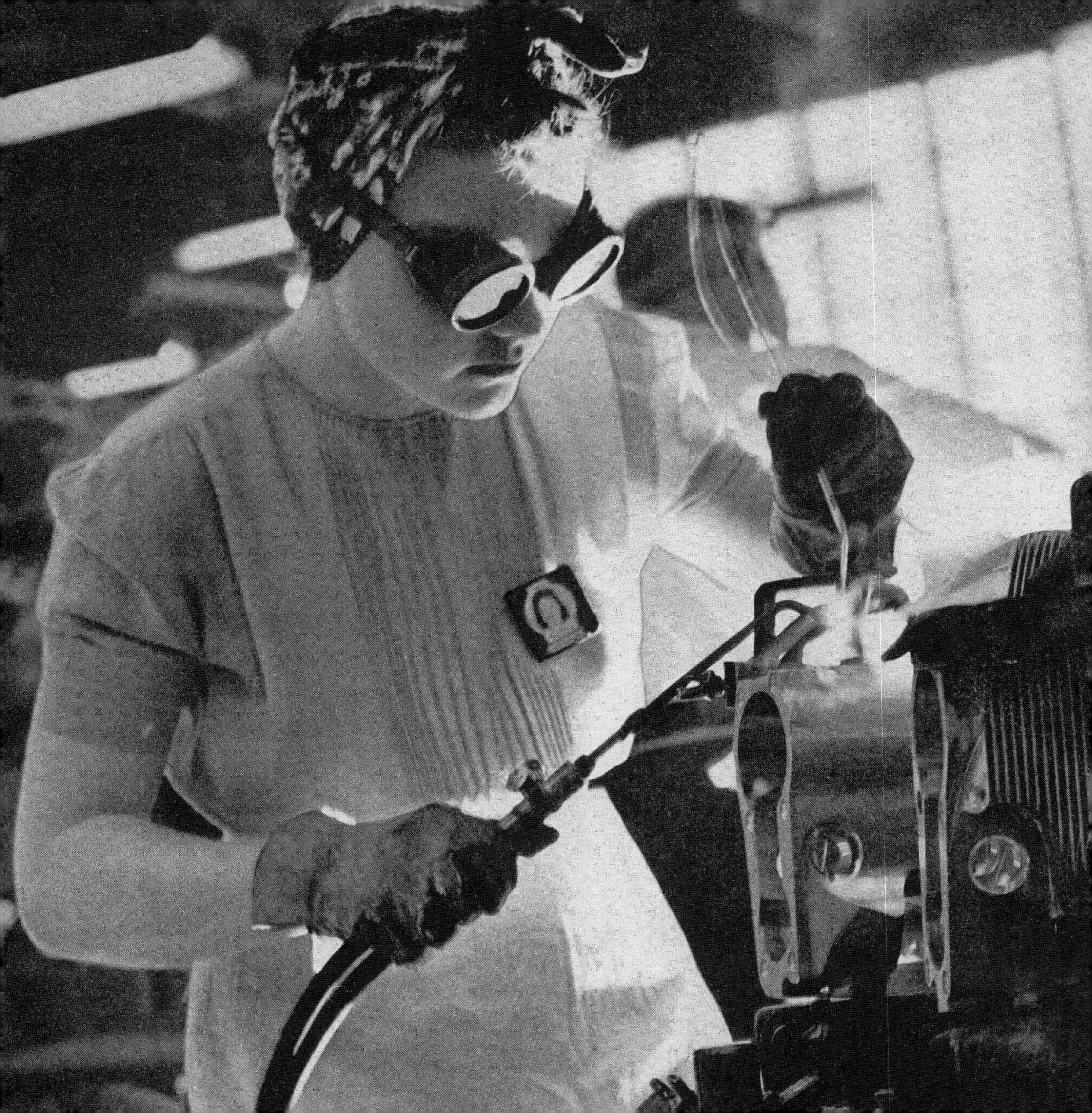

Female Inspectors

Vought–Sikorsky Aircraft was best known as a manufacturer of high-performance military and commercial airplanes. Until a month before this photograph was taken the Stratford plant site had served as the supplier of the deflectors used on all Pratt & Whitney engines built by the parent company as well as by its licensees. A deflector was a small, formed piece of sheet aluminum alloy, which, when attached to the engine cylinders, directed and controlled the flow of air required to cool and equalize the heat on the cylinders' finned surfaces.

THIS INSPECTOR IS SEEN CHECKING A DEFLECTOR PART AGAINST A PRATT & WHITNEY ENGINE CYLINDER.

The deflector department—as well as approximately one-half of the sheet metal assembly department—was moved to a new branch plant. This plant, located on South Avenue in Bridgeport, was completely renovated prior to its occupancy.

The problem of moving the large quantity and variety of machinery into the new plant with the minimum interruption of production required careful study and planning. Drill presses, tool grinders, rivet squeezers, lathes, presses and spot welding machines, to name just a few, had to be moved through heavily travelled streets a distance of approximately six miles. Notwithstanding, the complete transfer was made with a loss of only one-and-a-half days of production—and, to top it off, the deflector department set an all-time record for production during the same month in which this changeover occurred!

1942

A Navy Wife Gets a War Job

Mrs. Janet Fromson, an applicant for the training program, steps off the bus for the first time in front of the Pratt & Whitney Aircraft plant. Miss Jane Edgar (top right photo), in the personnel building, gives her a friendly reception and an application blank. During an interview it is determined Mrs. Fromson has the stuff of which a competent worker is made. Following the interview, she provides a detailed medical history and is given a physical examination that records her present physical state. Mrs. Lela Beeson (bottom right photo) helps with the welter of paperwork. Mrs. Fromson fills out an investigation questionnaire, signs an espionage oath and invention waiver, submits a birth certificate, and takes out group insurance. The investigation section issues a badge and takes her fingerprints. The following morning, Mrs. Fromson reports to the training school to begin learning the operation of the etchograph. Her husband, Leo Fromson, a naval flight cadet, will pilot a craft powered by an engine that she helped to build. His pretty, blue-eyed wife is in training to help her country supply him and his fellow American fighters with the implements of war.

1943

A Modest Girl Hears from a Busy Hero Husband

The news that Lieutenant Orion C. Shockley had been decorated three times for heroism in the North African campaign might never have gotten around if it had depended on the Shockleys. Lieutenant Shockley is the type of soldier who, in the midst of a letter to his wife, may sum up a momentous experience with the casual remark, "By the way, I got a medal today."

His wife, Mrs. Helen Shockley, one of the first employees of Pratt & Whitney Aircraft, works at the employment office and is just as modest about her husband's exploits. She knew last July that Lieut. Shockley had been awarded the Soldier's Medal for heroism, but it was months later before people who work with her heard the news. A fellow employee made his routine inquiry about whether her husband had reported anything interesting lately. Nothing she could talk about, she answered, except that he'd been awarded a medal. Pressed for details, she revealed the medal—the Silver Star for gallantry—wasn't his first. Then she told about his Soldier's Medal. A day later, Mrs. Shockley received her husband's third medal, the Purple Heart.

When the news spread around the building, Mrs. Shockley was surprised at the excitement it aroused. "Heroism is part of the day's work to Orion," she said. "He doesn't like to brag about it, so I've been treating it the same way."

1943

Helen Shockley

Assembly

Miss Rosalie Grim (foreground) and Mrs. Fern Karner (background) inspected the parts of the power section laid out on the special individualized racks introduced by Pratt & Whitney assembly supervision.

1943

The People in Personnel

Each day, many interviews were conducted over the desk of the employment department's Mrs. Fern Bishop. Here Mrs. Bishop went over Mrs. Helen Hemmingsen's application blank. Mrs. Hemmingsen had decided she'd like to roll up her sleeves and get her hands dirty building aircraft engines. She subsequently got her wish. Mrs. Hemmingsen was at the training school, a member of the women's machine operation classes. Living up to the promise extended by the results of her aptitude examination, Mrs. Hemmingsen soon becomes a first-rate machine operator.

1943

Fern Bishop (Left) and
Helen Hemmingsen (Right)

To help workers meet housing and transportation difficulties, the Pratt & Whitney personnel department set up a force of experts. A new employee was informed of the revised Pratt & Whitney employee programs for transportation and housing support. For transportation, fellow passengers could be found for a driver who then qualified for supplemental gasoline for driving a car to work. If seeking a ride, an expert in the department could find one. And if looking for new living quarters, the new employee could apply for support to the housing expert.

In the top right photo, Mrs. Myrtle Brewer, group leader of employee files and applications, explained the filing procedure to Miss Bertha Hicks. Behind them is Miss Doris O'Dell.

Myrtle Brewer (Left), Bertha Hicks (Right), Doris O'Dell (Background)

In the bottom photo, new employees find rides as Annabelle Kelly, Marcella Chronister, and Lorraine Barham (seated, left to right) check their files for drivers who lived in the neighborhood of Elnora Clark (left) and Barbara Boyd, on their first day at work.

Colorful Backgrounds in Every Position

After spending a year as resident physician in the base hospital on the 16-million-acre Navajo Indian reservation in northern Arizona, Dr. Margaret L. Dale thought she would be well prepared for any contingency that may arise on her new job as staff physician in the Pratt & Whitney medical department.

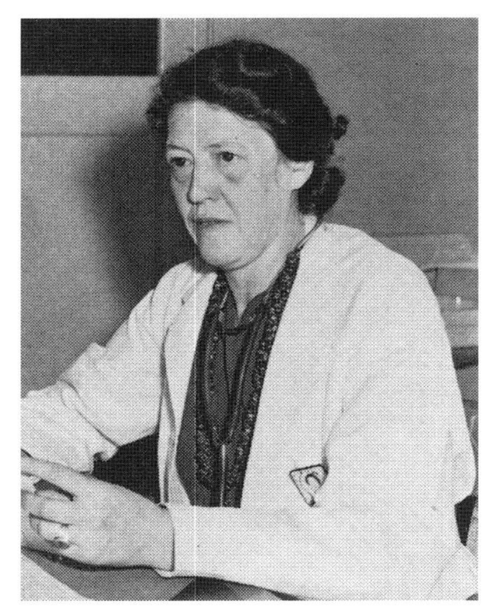

MARGARET L. DALE

Kitchen Science

Mrs. Mary Mykland, Pratt & Whitney's dietitian, wishes they'd added a course in crystal-gazing to the home economics curriculum at the University of Kansas, where she studied, because it would take just about that to figure exactly what the demand would be at the cafeteria on a given day for Boston cream pie, tomato salad or macaroni and cheese.

Such speculation is only one little detail of her job. Her mornings were a race around the net of kitchens to supervise food production. At serving time, she observed employee tastes. If on a day when chili was served, for instance, salad consumption went into a slump, she made a mental note, and on the next chili day, fewer salads would be made.

The afternoons she spent doing her office work of making up menus and recipes. As in any household, her planning revolved around the ration stamps available as well as the vitamins needed. It took a lot of figuring to get the points to come out even with the Pratt & Whitney appetite.

1943

MARY MYKLAND

On the Go

The entire office and factory floor was covered many times a day on foot by the four personnel groups represented by these women. Four of Pratt & Whitney's full-time pedestrians were Mrs. Harriet Bolar, a monitor; Miss Nadine Ditto, an escort; Mrs. Carita Zimmerman, a mail route clerk; and Mrs. Jean Butcher, a time-keeping messenger.

In the factory the mail was delivered by tricycle. The uniformed mail route clerks shown here were Miss June Willey, Miss Rosanna LaGue and Miss June Walton. Miss Willey and Miss Walton were known along their routes as "big June" and "little June," as Miss Willey, who was five feet tall, had a four-and-a-half-inch edge over tiny Miss Walton. Each cyclist covered a 2/3-mile route 26 times a day, the girls taking the clearest aisles in the factory area. In more congested locations, where navigation was difficult, Charles Shepard and Gene MacCready made the rounds. On the outside route, over which mail was carried to the personnel and construction buildings and to the boiler house and other outlying structures, Joseph Trotter delivered the mail. Eventually, the tricycle force would be increased to about six times its present size.

1943

Walking, Left to Right: Harriet Bolar, Nadine Ditto, Carita Zimmerman and Jean Butcher

Bicycling, Left to Right: June Willey, Rosanna LaGue and June Walton

Driven to Succeed

When a newspaper reporter in Topeka got an eyeful of Miss Mary Ellen Soden, who, in her chauffeuse (female chauffeur) uniform, had driven Pratt & Whitney executives to dine with Governor Schoeppel, he thought she should be written up in the Topeka Daily Capital. "It will be a sad day," he concluded, "when there are again such things as men chauffeurs." The seven chauffeuses are shown here in their new summer garb. The Pratt & Whitney uniform, he asserted, was "the best uniform of all."

1943

Left to right: top row, Miss Dorothy Miller and Miss Janet Mitchell; second row, Miss Bessie Blackburn, Miss Martha Whitaker and Miss Soden; bottom row, Miss Ellen Markle and Miss Pat Oakes

Dressing for Success

Miss Montez Dusenberry, 1941 Queen of the American Royal, modeled the first escort uniform of light olive drab gabardine. Her associate escorts, Miss Joan Royal, former fashion designer; Miss Maurita Eggleson, once Chillicothe Homecoming Queen; and Miss Betty Walker, a Cessna Aircraft beauty contest winner, examined the uniform that they would wear. The material—rayon and "air-lac"—was crease-resistant and suitable for year-round wear. Front and back kick pleats in the skirt were stitched down so they wouldn't lose their press, and the emblem on the sleeve was embroidered in gold. With the uniforms the girls would wear identical British tan shoes.

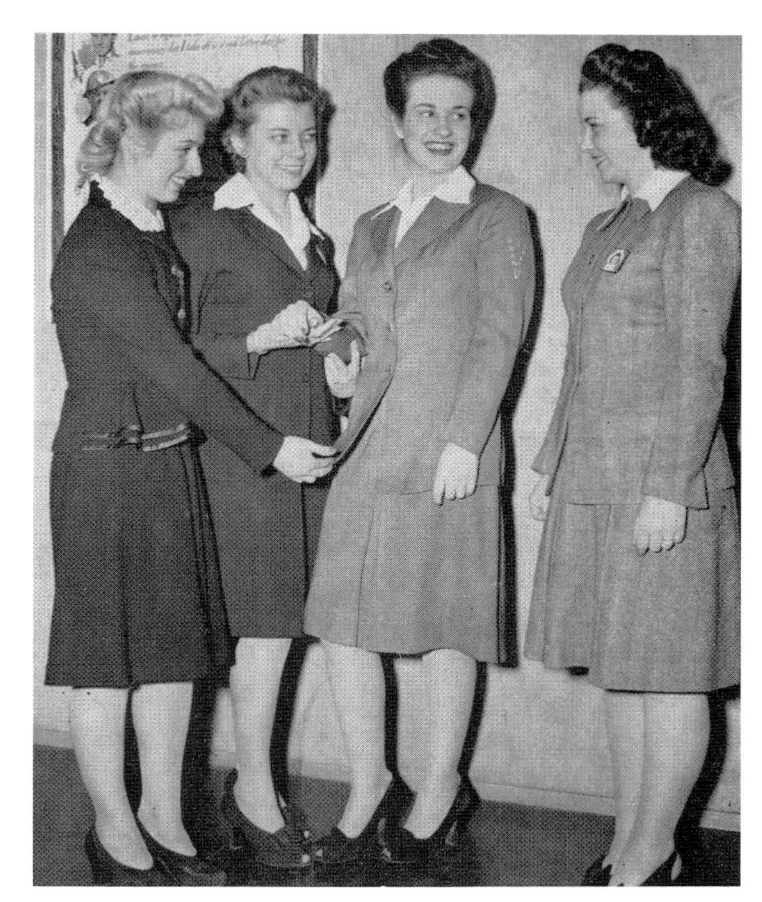

Left to Right: Joan Royal, Maurita Eggleson, Montez Dusenberry, and Betty Walker

1943

Left: Annie Laurie Warehime, former teacher and licensed pilot, was the chief receptionist at the Pratt & Whitney Aircraft plant.

Right: Ruth Groves prepared to go to town in her new courier's uniform of brown whipcord.

Timekeeping Class

A potential Pratt & Whitney timekeeper had to discard many notions of time she learned in grade school. Elements of time were not what they seemed. What appeared to be an hour may be an hour and a half or even two hours in timekeeping arithmetic. A minute may become six, or five minutes may vanish completely. There were occasions in timekeeping when six and a half equaled eight. To add to the confusion, timekeepers told time by two methods: One dividing the day into conventional twelve-hour segments, the other considering it a twenty-four-hour unit.

An industrial timekeepers' school with a regular course of study was something of an innovation. Previously, time clerks were needed one at a time. As in any clerical job, the timekeeper learned at her post with an experienced worker. But the demand from the Pratt & Whitney factory floor for timekeepers had necessitated their mass production.

A course of study had to be devised for the school and a textbook compiled. Fortunately, Frank A. Clifford, chief timekeeper, and his assistant, E. Ray Edwards, were experienced educators. Questions and examinations were concocted. A reference manual was compiled, and as in a regular classroom, a schedule of daily practice periods, recitations and tests was established.

1943

After all the employees punched the clock at the beginning of the shifts, Marjorie Kiloh removed the cards from the rack to make the daily extension of time.

INTERVIEW

MARY LORA WOLFE, DOLLIE LUCILLE CHAPPELL, BONNIE FAYE BROCKMAN

Mary Lora (Reed) Wolfe, her sister Dollie Lucille (Reed) Chappell, and their friend Bonnie Faye (Wolfe) Brockman worked first at North American Aviation during the war, and then were told to move to Pratt & Whitney as the war came to an end. They all worked the evening shift. Bonnie worked in the machine shop drilling small oil holes into parts; Lora and Dollie worked in a different department. After the jobs came to an end, it was hard to give up the $1.10 per hour. Most of the money earned was sent home to care for their families.

Mary Lora (Reed) Wolfe, Dollie Lucille (Reed) Chappell and Bonnie Faye (Wolfe) Brockman (in unknown order) in their work uniforms

A Natural for Teachers

LAWRENCE FLAGLER AND WALTER STANLEY GIVE THEIR JOB TICKETS TO MRS. MARIE STODDARD, THEIR TIMEKEEPER. MISS CARMON JENKINS IS A TIMEKEEPING TRAINEE.

Schoolteachers were especially apt timekeepers because they were accustomed to schedules and routines. Some disciplinary ability, too, was required of them, where again their teaching experience came in handy.

For instance, while the timekeepers' problems usually center around the balancing of reports, and not around the behavior of the clock punchers, some timekeepers said the noon rush for the clock was just a little reminiscent of the impatient elbowing among the schoolchildren at morning recess time. There was one important difference: the eager ones here were bigger.

In the tool development room one day, a clerk punched the clock at 11:10, and the hungry tool roomers, assuming it was lunch time, made for the clock en masse. It was a big job for a tiny time clerk like Mrs. Vernabelle Schmidt to get such burly charges back to the fold in time to avert the disaster of an early ring-out, but that was one of her responsibilities when the occasion arose.

1943

BETTY LEE JONES EXPLAINS A PROBLEM IN CALCULATION OF WORKING HOURS TO A CLASS OF TIMEKEEPING TRAINEES.

Class in Session

Working on gauges with graduated diameters, this class of girl trainees learned to read micrometers. Their work would be checked by the instructor until they had successfully made accurate measurements. One textbook figured that there were 16 basic machine tools. In the single-purpose operator function, a person was taught to do one thing well in a short time. But in the case of the United Aircraft Corporation training program, the theory was not the single-purpose operator, but the single machine tool operator. Thus, a man who was being trained as a milling machine operator did not learn only one type of milling machine; he learned all of the seven various models.

1943

Women studying micrometers

Scenes from the Training School

Completed parts, made in the training school, were tagged for identification.

Left to Right:
Margaret Fristo,
Rose Graham and
Catherine Malone

All employees were urged to cooperate with the rationing boards in granting extra mileage rations, by following the example of these young women, who had formed a group riding club.

Left to Right:
Katherine Johnson,
Betty Leach, Arlene Beckford,
Ione Neustadt and
Betty Lotz

People came from all over to work at Pratt & Whitney. These women came from Hartford to take top secretarial posts here.

Left to Right:
Laura Nelson,
Beatrice Moore and
Margaret Lamprecht

In the plant protection department, women were being trained to take over the job of fingerprinting new employees. D. G. Hilton, a guard, instructed Irene Ward in this procedure. H. E. Van Campen, in charge of Magnaflux at the training school, was being fingerprinted.

Left to Right:
D. G. Hilton, H. E.
Van Campen and
Irene Ward

Receptionists Looking Sharp

Here the staff of receptionists donned their new uniforms and moved into the first permanent lobby in the administration building. Miss Annie Laurie Warehime, chief receptionist, is standing, reading roll call.

1943

Left to right: The escorts are Murita Eggleson, LaDean Holland, Eleanor Laizure, Montez Dusenberry, Susanne Barton, Olive Good, Nadine Ditto, Lillian Riling and Joan Royal.

Mrs. Fix-It

Workers shouldn't have been surprised if their call to the office machine repair room for typewriter services brought a lady repairman. Mrs. Eloise Bruce came to Pratt & Whitney as a stenographer. After previously pounding a typewriter for seven years without knowing what made it run, she decided to join the office machine service staff. "The training course had been a little like a game of hide-and-seek," Mrs. Bruce said. Her supervisor would send her away from a perfect typewriter into the hall. Then he would gum up the works in some way—put the ribbon on backwards, remove a part, or juggle the line spacer so the paper would slip. Mrs. Bruce would then be summoned into the room to see if she could discover what was missing. The accomplishment Mrs. Bruce was proudest of was that she made tools of her own. Spring hooks, tiny instruments for making adjustments in the minute crevices of a typewriter's mechanism, were unobtainable, so Mrs. Bruce set to work on the repair shop's lathe and made one. Soon Mrs. Bruce was out on her own on service calls, taking apart, cleaning and putting together many different machines.

1943

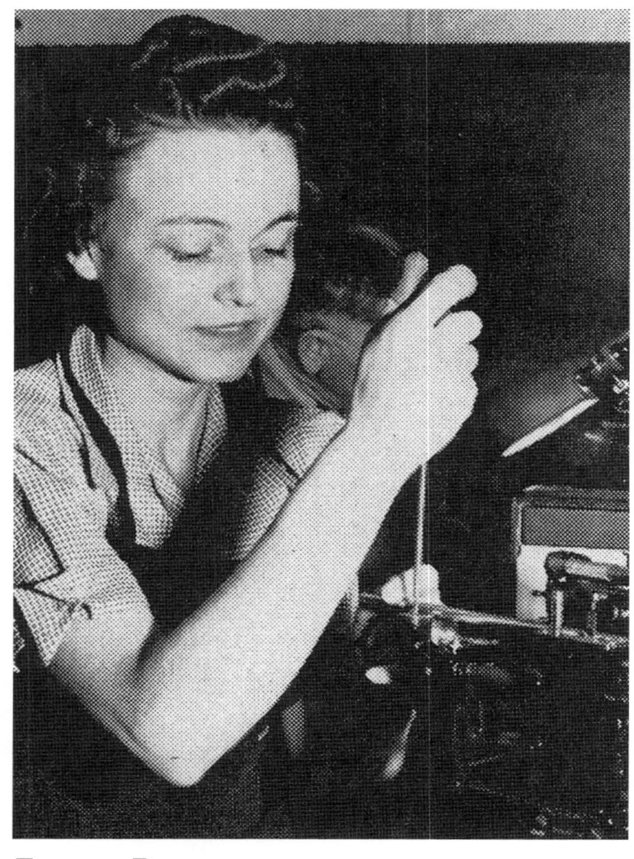

Eloise Bruce

A Psychologist Turned Inspector

Miss Margaret E. McCaul had to admit that she probably did a bit of a selling job on her own subconscious. "It was a case similar to the story of the man who went around telling everyone that he should go down to see the big elephant which has just arrived with the circus," she said. "He hadn't seen the elephant himself, but after he had told everyone about it and had sent everyone scampering down to see it, he got curious and had to go take a look himself."

As a special guidance and personnel supervisor at Westport High School, Miss McCaul had been advising many students to hitch their wagons to the ascending star of the aircraft industry. The more she told them of the industry's wonderful future and of its importance in winning the war, the more intrigued she became with the idea of finding a place in it herself. She found that place—as an inspection trainee on the first shift at the school.

For someone with a master's degree from Columbia University and research work at Stanford University, taking elementary college math courses at the training school might have seemed incongruous. Miss McCaul admitted that despite the fact that she majored in mathematics in her work toward a bachelor's degree, she had to do plenty of "boning up" on her homework.

1943

MARGARET E. MCCAUL

INTERVIEW

SANTINA BRANCATO CATALINA

Santina Brancato Catalina worked at the Pratt & Whitney plant in Kansas City. Santina's family owned Fairyland Park, an amusement park in Kansas City. Santina worked at both Pratt & Whitney and also Fairyland during World War II. During the war Fairyland converted some of the rides to bomber planes. All of the "Rosies" from Kansas City went to Fairyland Park when they were off work. It was the happening place during that time period!

SANTINA BRANCATO CATALINA

1943

FAIRYLAND AMUSEMENT PARK

One of many Pratt & Whitney women's teams in 1943

Fun in the Sun

The water's fine, but the July sun was a little strong for Pat Johnson, production engineering. Pat Angelbeck, mailing department, protected her back with sunburn oil as Marie Holdren, mailing department, and Jean Wilcox, stenographic pool, looked on.

Newsworthy article from 1943

The Owl Shift Has a Fling

Wartimes always upset routines. Some Pratt & Whitney Aircraft trainees went to work at midnight and finished their day at 7 o'clock in the morning. Likewise, a bowling match at 8 o'clock was no different from a normal-hours man who rushed off for a golf game in the afternoon. Miss Jean Gregory (shown here) was willing to expend pounds of energy trying to knock down a few tenpins.

1943

The Family That Works Together...

Pratt & Whitney Aircraft drew members of many families closer together by the interests of a common working place, as the Berlin and Breshears families could attest.

W.B. Berlin was one of Pratt & Whitney's first Treasury employees. Mrs. Berlin, on applying at the plant, found that she could take the inspection training course and still be home before her 12-year-old son, Dean, arrived from school. She also had time to have dinner well under way when her husband came home from the plant. Their daughter Jeanne returned from Kansas State Teachers college at Pittsburg and started the inspection course as well. Both Mrs. Mae Berlin and Jeanne were in purchase parts inspection, Mrs. Berlin inspecting piston rings and Jeanne inspecting conveyors.

When Wheatland, Missouri's Merdith Breshears went into the Army Signal Corps, his brother, Malvern, moved to Kansas City with his wife, Verna. Shortly thereafter, Malvern's father and mother, Ira and Fay Breshears, followed. Ira was placed in the materials division, and Fay was assigned to machine operator training. When Malvern and his mother worked on the same machine, one trainee occasionally called another "Mom."

1943

Malvern Breshears (Far Right) explained the operation of the machine on which he worked to (Left to Right) his father, Ira Breshears; his wife, Verna Breshears; and his mother, Fay Breshears.

Grandmother Runs Turret Lathe

Mrs. Pearl Hasten is hard at work learning the operation of the turret lathe and other huge precision machines as a member of the initial class of women production trainees at the Pratt & Whitney Aircraft training school. When Mrs. Hasten saw the courageous devotion to duty displayed by her soldier son when he was ordered away from his wife and infant daughter to an isolated post in Alaska, she decided that she had a war job to do, too, and enrolled for machine shop instruction at a government-sponsored War Training school.

Her son had returned to the mainland to enter officers' training school. He wrote of his pride in her, saying: "You are doing just as much to win the war as any of us." And that was the spirit that bright-eyed, brown-haired, youthful Grandmother Hasten and the women with her were taking into their jobs.

As W. F. Grier, director of the school, said: "All our women trainees are working with an eagerness and enthusiasm that can't be beat. And we are especially pleased at the way this first class of women production trainees are getting along. Their grasp of blueprint reading and machine operation is remarkably good, and in their work on the machines they are meeting the requirements established by set-up men and male operators."

1943

Pearl Hasten, first grandmother to enter the Pratt & Whitney Aircraft training school, "miking" an engine part.

A Calm, Steady and Helping Hand

When Gladys Russell, first-shift training school counselor, told trainees how to hold down a job and a home at the same time, she knew her advice was good. She'd tried it out. While she accumulated years of experience as a machine operator, she has raised an 18-year-old daughter.

If discouraged workers complain to her that blueprint reading and precision machine math aren't easy, she knows what they mean. When she started to work at Pratt & Whitney, it was as a production trainee, and she had her troubles. Keeping up with the class took her until midnight sometimes.

When the company was looking for a trainee who liked people and understood their worries, she fit the description. She became a counselor, but she wasn't through studying. The trainees' problems ranged from how to sew on buttons to how to compute income tax. So she was awake nights again, boning up on tax legislation.

When told to the calm, sensible Russell, the trainees' mountainous anxieties usually shrank down to the molehills they really were. A quiet talk with her could accomplish anything from balancing a budget to quelling a fear for a serviceman abroad or a sick child at home.

1943

GLADYS RUSSELL

Females in the Factory

Mrs. Mildred Brock and Mrs. Helen Elliott were the first feminine employees in the factory proper. They had received their preliminary instruction at the training school, and were being further trained at the plant to be permanent employees of the tool room and supply department. As crib tenders, they would learn to read precision instruments down to one ten-thousandth of an inch. They also would be charged with the responsibility of checking tools in and out, which required familiarity with many hundreds of tools.

1943

Mildred Brock (left) and Helen Elliott (right)

That Fabulous Eye: X-Ray

Alice Miller adjusted a part for X-ray, centering the big tube apparatus directly over the part. If there were cracks, gas bubbles or slag in the metal, these would absorb the rays in a pattern that differed from that of a perfect part.

In developing the film, Lois Young uses a process similar to that used in the ordinary photographic darkroom. As she took the developed film from its final wash, she turned on the darkroom illuminator to make a quick examination of the wet radiograph. Satisfied with the work, she placed the radiograph in the drying cabinet.

1943

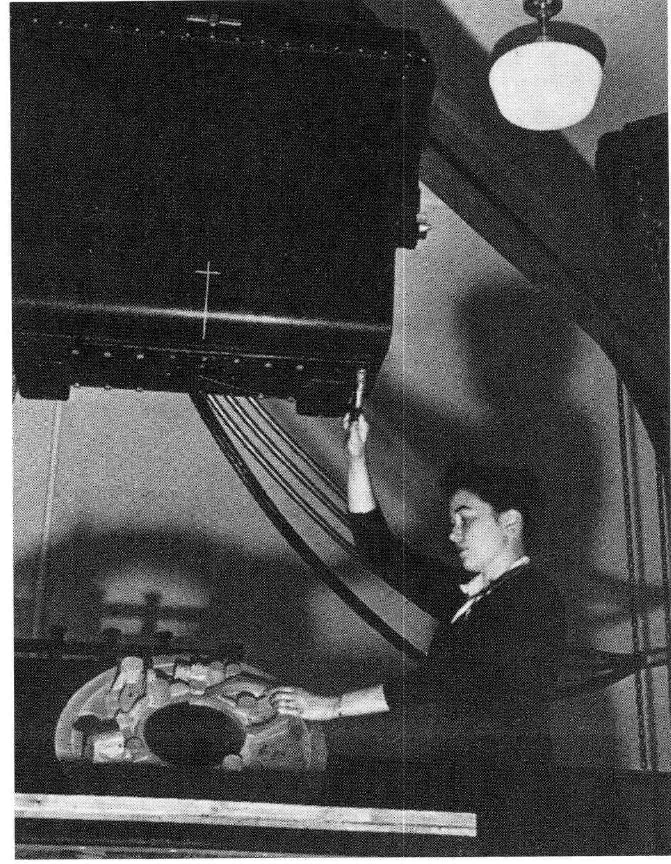

Alice Miller

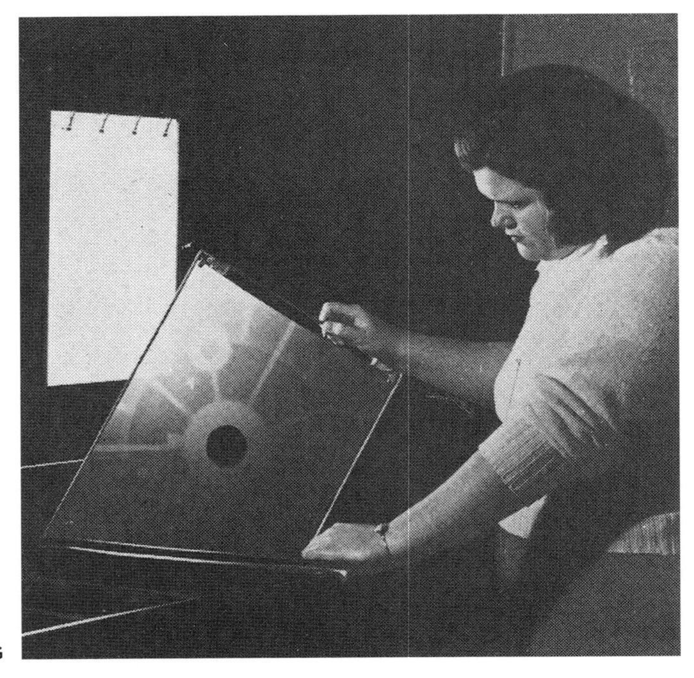

Lois Young

Good Girls and Bad Habits

Miss Marguerite Toplikar remembered that her instructor said she should keep her machine wiped off. So while she waited for the grinder to do its work on a part, she started dusting. The moving machine snatched the cloth away from her and spun it around the wheel. She was lucky. It might have snatched her hand, too.

Miss Dorothy McVickers was similarly conscientious about keeping the chips off her lathe. She dove in with both hands. The blue tinge in the metal shavings came from an intense heat. The shavings had sharp edges. Next time she'd remember to remove them with a chip hook.

1943

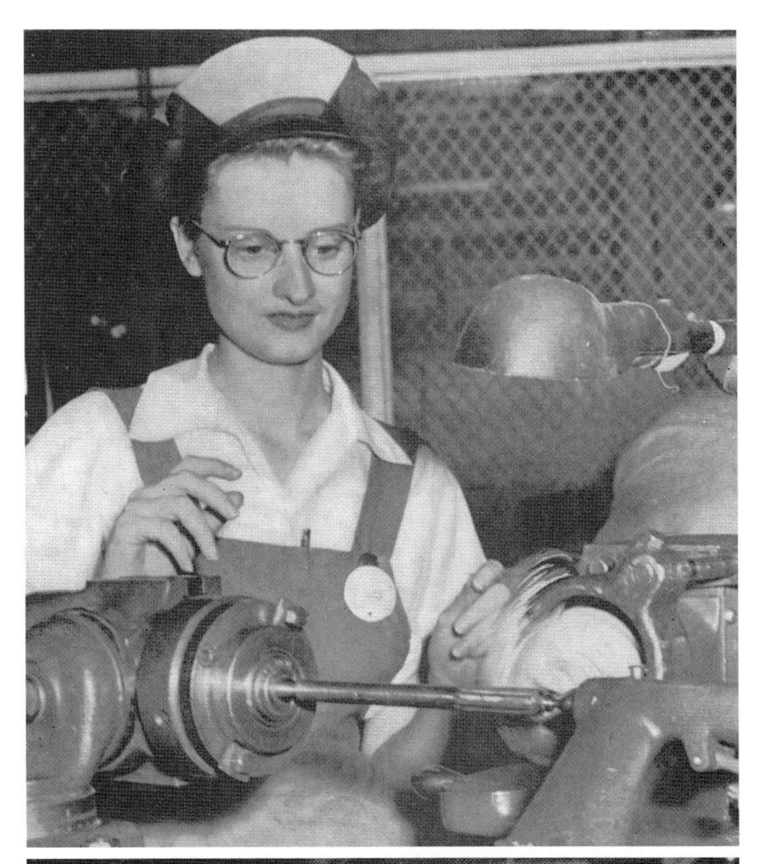

Top: Marguerite Toplikar
Bottom: Dorothy McVickers

INTERVIEW

ROSE PIXLEY

I began my employment at Pratt & Whitney in 1942 and worked there until the end of World War II in 1945. After completing high school in Lincoln, New Hampshire, my friend Cecilia and I decided to move to Connecticut. Jobs were plentiful and the pay was sufficient there. My role at Pratt & Whitney was to secure rivets and spark plugs on war planes. I worked on an assembly line and to complete my work used a Riveter gun. Being a woman at Pratt & Whitney enabled women like Cecilia and myself to contribute to the wartime efforts. I enjoyed my time at Pratt & Whitney and was proud to be able to contribute. I have many fond memories of my time at the company. We knew that we were helping out during the war the best we could. It was a special kind of community of women at Pratt & Whitney, we were like a big family. I saw a change in responsibilities for women, as well as a change in our view of our role in the war. It sure was an experience for a "little country girl."

1943

ROSE PIXLEY IN 1942 IN EAST HARTFORD, CONNECTICUT

Preparing a Wasp Engine for Shipment

Pratt & Whitney Aircraft and Hamilton Standard Propellers divisions of United Aircraft Corporation of East Hartford provided women with a variety of jobs. Here Ann Prior prepared a Twin Wasp engine for shipment.

1943

Left to right: Governor Raymond E. Baldwin, P&WA General Manager H. M. Horner, Lieutenant General Hugh A. Drum (Commander of the Eastern Defense Command and First Army), Ann Prior

Prize Awarded to First Woman to Suggest Machine Tool Improvement

Natalie B. Conradi, a native of Russia who trained at the Pratt & Whitney Aircraft Training School and has been with Pratt & Whitney Aircraft for only three months, was the first woman to offer an accepted suggestion—that the spacer on the Van Norman milling machine she operated be permanently attached. She accepted her prize of $5 in War Savings Stamps from Assistant General Manager William P. Gwinn as Factory Manager George H. D. Miller looked on.

Conradi's award typified the greatly increased interest shown by all Pratt & Whitney Aircraft employees in making suggestions beneficial to their war jobs. During the first 10 months of 1942, employees made 2107 suggestions concerning increased production, safety and morale, of which 394 were found acceptable. Recognizing the importance of these suggestions to the war effort, Pratt & Whitney Aircraft increased the aggregate semi-annual suggestion awards from $275 to $1425, and the total monthly prizes from $50 to $260.

1943

Left to right: Natalie B. Conradi, William P. Gwinn and George H. D. Miller

GIRL-
AND A GEAR
NOVEMBER
1944

Which Are the Prettiest?

A fellow came out of the factory with a suggestion for us. Said his department had the prettiest girls in the plant. Why didn't we ever take their picture? We don't say these are the plant's most handsome girls, or the most handsome in their departments. We are not beauty experts. But these do make a nice group of pictures. What do you think?

1944

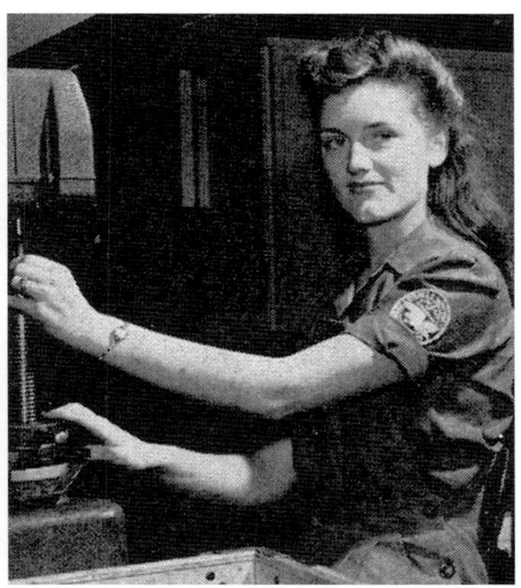

Top: Ellen Polfer, Department 784;
Virginia Ullom, Production Engineering;
Middle: Lorene Hermon, Department 764;
Geraldine Guthrie, Department 777;
Bottom: Betty Brinker, Department 710

Good Girls Bad Habits II: The Office

Miss Ellen Williams, inspection, juggles two phones.

Miss Pat Story, badge control, slips out of her shoes

Miss Lee Kline, stenographer, types calmly while her manager waits impatiently behind her.

1944

Miss Suzanne Casey, plant engineering, files with her feet in a desk drawer.

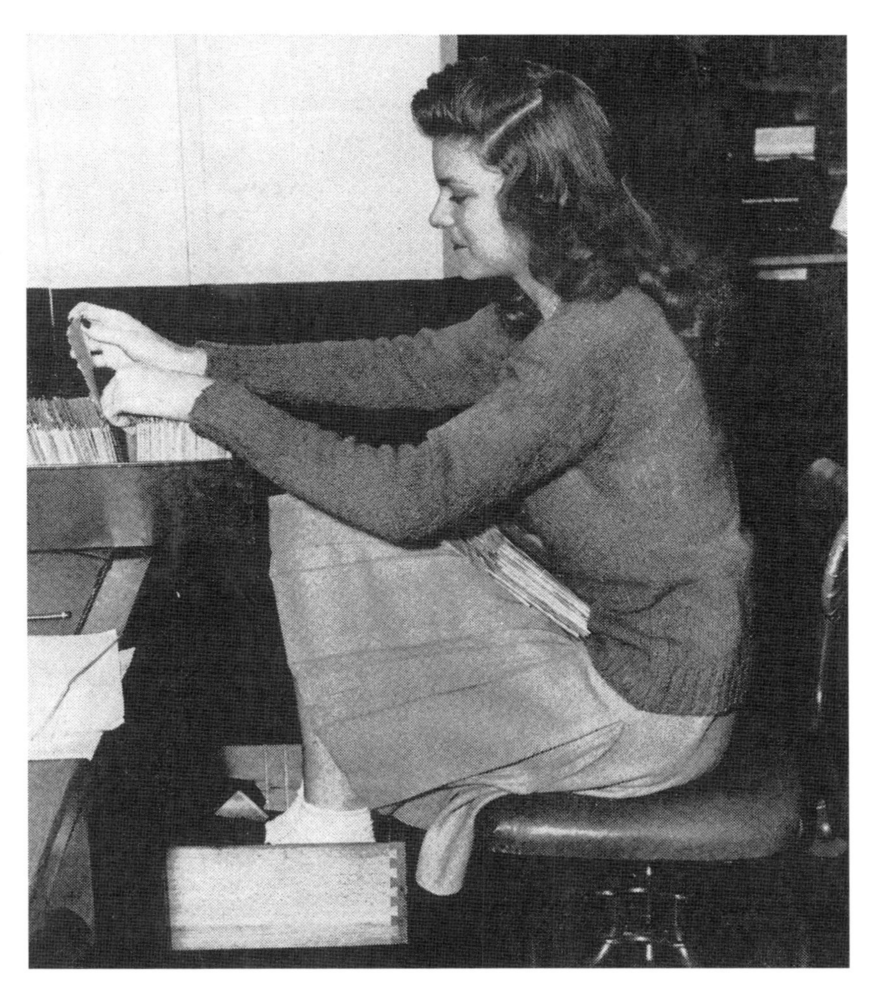

Miss Peggy Tellman, absentee tabulating, has a misplaced pencil behind her ear.

1944

Odds & Ends

More House Hunting

Looking for a house is easy at the transportation department. Mrs. Agnes Sweeney (second from right) shows home shoppers what was available in her housing files, while Pearl Graham and Betty Rice got new real estate tips on the phone.

A Dog's Life

A dog's life is good at Pratt & Whitney. Fannie, the plant's earliest inhabitant, basked in the burlap yard of the home built by the plant carpenters for her and her six new pups. Flora Smith is shown holding the pup she plans to adopt.

Spoils of War

Alice Lowe examined the souvenirs of battle sent home from Germany by her brother Pfc. Albert Evans.

1944

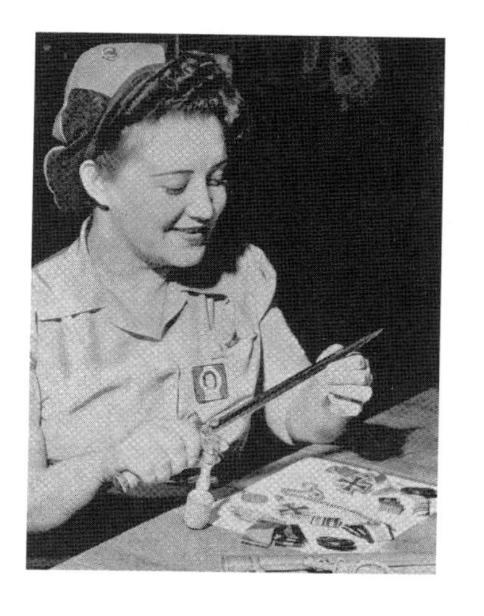

Car Pool Girls

Mrs. Bernice Anderson took seriously the rule that a courier must be ready for any emergency. The volunteer blood donor she was driving from the plant didn't have the right type of blood, so she rushed in and gave the transfusion herself. "We never know what will come up on a trip," said Ruth Dupy. "We have to do things almost as fast as a Hellcat squadron," one of the drivers said. The drivers all had passed a rigid safety department driving test to qualify for a Pratt & Whitney driver's license. The girls had to get along with the many types of persons who rode in their cars.

They were a varied group. Tall, blonde Dorothy Miller was a fashion model and a department store supervisor before she joined the car pool. Tiny, dark Martha Gipson was a chemical laboratory assistant. Velma Lamb was married with three children. The driving job was Martha Phillips' first. Ann Smith held a clerical job. Janet Mitchell, supervisor of the car pool, formerly taught vocal music and English in Missouri high schools.

The handsome, brass-buttoned uniforms the drivers wore in winter aroused a lot of curiosity. All the girls have heard many times the question, "What branch of the service are you in?" They have also been mistaken for telegram couriers, elevator starters and even FBI agents.

1944

Standing, Janet Mitchell, Supervisor
Front row, right to left: Martha Gipson and Martha Phillips
Back row, right to left: Ruth Dupy, Bernice Anderson, Ann Smith, Velma Lamb, Dorothy Miller and Dorothy Winger

Women at Play

The Pratt & Whitney girls' basketball team has been in first place in the women's division of the War Industries league all year. The squad was as follows: Front row: Betty Giles, La Dean Robertson, Frances Hornbuckle, Sally Pack and Billie Sherrill. Back row: Darlene Powell, Evelyn Miller, Vivian Maxey, Johnnie Bowles, Mary Ann Leek and Norma De Vorss. Merle Steele was the coach.

The girls' glee club practiced under the guidance of Rachel George, director. They were Wanda Hargrove, Lenora Blum, Helen Alcorn, Velma Arney, Gladys Thompson, Frances Weathers, Wilma Marsteller, Jerdie Farris, Marie Searcy, Carlie Myers, Bernadine Nigh and Gladys Russell.

1944

A Former Athlete Has All the Strength She Needs

Mrs. Jennie Pepper wasn't perturbed at all when the foreman said: "You'll have to have strong arms and hands to do this job." She said if she couldn't do a good job of grinding knuckle pins she didn't know who could. The strength of Mrs. Pepper's arms, in fact, was sufficient to have brought her through three state golf tournaments as top winner. She won the Missouri championship in 1941. But the war and the end of tournaments meant golf soon took up less of her time. With more time on her hands, she applied for a job at Pratt & Whitney. Her two sons were fliers, one in Italy and one in training.

1944

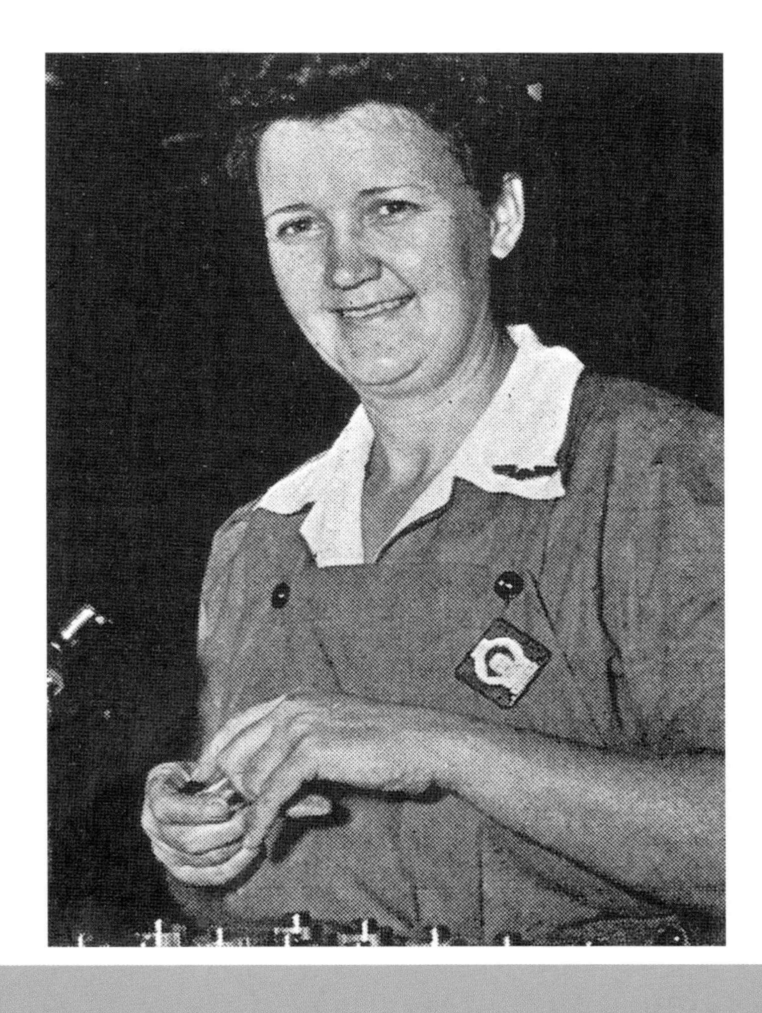

Concentrated Course: A Trainee Tonic

Stepped-up production at the Pratt & Whitney Aircraft Missouri plant had necessitated a corresponding speed up in the training of new workers. Peacetime apprenticeships of several years had already been compressed into an intensive training schedule of a few months. The Pratt & Whitney training school was turning out skilled inspectors, machine operators and engine assemblers in six weeks.

1944

Foreground: Norma Jensen pitches in with cleaning.

Left to right: Unidentified man, Vinitta Gross, Helen Temple and Dorothy Gillum

Junior Instructors

Five women trainees put on white coats after finishing their training school courses and went to work in the shop and classrooms as junior instructors. Standing among the highest in their respective classes and having backgrounds and personalities that suited them for the work, they were the first women to be placed on the school faculty. Valda Gilmore, an inspection instructor, had a degree in education from the Kansas State Teachers College at Emporia. Edna Warren, a junior blueprint instructor on the third shift, also had a degree in education. She was a teacher for 10 years. Hulda Martin, a graduate in psychology from the University of Pittsburg, worked in the Harvard University radio research laboratory until September 1943. Gladys Taylor, 21 years old, had only brief experience as a die-casting inspector after finishing high school in Glasgow, Missouri, when she entered the training school. She had always liked mathematics, and found the course at the school simple. Following training she served as a first-shift mathematics instructor. Ellen Westdal had trained to be a nurse and had never seen the inside of a machine shop before she entered the school. The women hadn't been at their jobs long, but they agreed they liked aircraft training instruction.

1944

Edna Warren

Left to right: Ellen Westdal, drill press; Gladys Taylor, mathematics; Valda Gilmore; and Hulda Martin, inspection.

Grandma Works to Win

Irene Cooper (top) was one among many hundreds of youthful Pratt & Whitney Aircraft Missouri grandparents. The youngest grandmother found by the photographer was machinist Bessie Acuff (center). Harriet Collier (bottom) was an inspector with 18 grandchildren. Ida Taylor, Sina Long and Ala Cummins (bottom, left to right) were three grandmothers who worked side by side each day; they averaged five grandchildren apiece.

1944

They Work to Bring "Him" Back Soon

Mabel Kennedy and Geraldine Guthrie, who worked together in the ring section of department 777, first shift, had a lot to talk about. Mostly, they were conjecturing about the date of the invasion and the probable length of the war. They talked a lot about the distance from Kansas City to "someplace in England." Both their husbands were there, stationed in the same camp. The two soldiers weren't acquainted, but a photograph of each one was on the way to the other. The wives were hoping their husbands would look each other up. Sometimes Alice Hedges, from the next inspection table, joined the get-together. Her husband was in England too, albeit in a different area.

1944

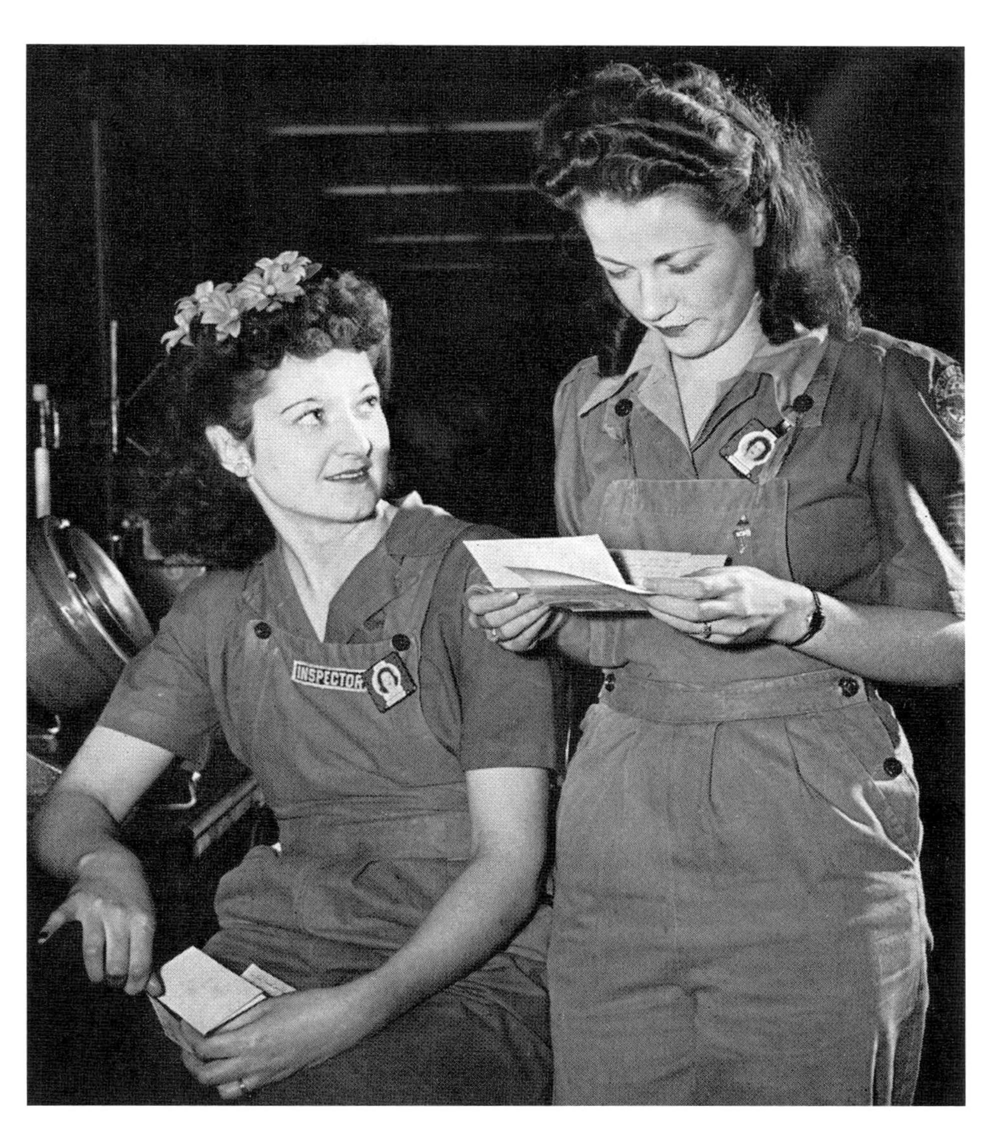

Mabel Kennedy and Alice Hedges compared letters from their husbands in England.

Evelyn Marie Welch worked while her husband was a waist gunner on a B-17.

INTERVIEW

BONNIE RUTLEDGE SHAW

Both of my parents (mom Audie Yocum) worked at Pratt & Whitney and after I graduated high school at 17 I went to work there also in Kansas City, Missouri, from May 1944–1945. I operated the lathe after the machine was set up. Mainly I removed burrs from machine parts. It was fun in a way and we felt we were helping the war effort. Women were accepted in the workplace for the first time in professions other than the traditional jobs of secretary, nurse and teacher. I made 80 cents an hour and I thought I was rich. I saved enough money working at Pratt & Whitney to attend Kansas City Business College after the war.

1944

Top: Bonnie Rutledge Shaw

Bottom: Audie Yocum

NORMA JEAN HENRY BOWERS

I worked at Pratt & Whitney – Kansas City, Missouri, from June 1944 to September 1945 as a tool sharpener. I was thrilled to get the job. I sharpened the drill bits and put the radius on them by hand on the grinding wheel. You had to have a very steady hand to put the radius on. I had a high success rate with very little waste. The tools I sharpened were used to build the Pratt & Whitney 2800 engine for the DC4. A lot of women worked there and I enjoyed the work. The men that were there were very appreciative of the women. I made good money and helped with the war effort.

It was very different work than that of my mother's generation. I was able to save enough money working at Pratt & Whitney to pay for airline school. After the war I got a job working for Capitol Airlines in Washington, DC. I started out at the ticket counter working different shifts. I asked for a transfer to work in the general offices (and) was hired in the general office as a parts buyer. One of the things that I purchased were the parts for the Pratt & Whitney 2800 engine for the DC4.

1944

THELMA M. HARALSON

Like her father, Walter L. Haralson, Thelma worked at Pratt & Whitney – Kansas City, Missouri, during World War II. Born in 1919, she began working at Pratt & Whitney in her twenties shortly after the bombing of Pearl Harbor.

Working on engine assembly, Thelma said there were some jobs she did that "were difficult for the men because it required working in small spaces with smaller hands to be able to tighten bolts, for instance. I felt very proud to work at P&W."

Thelma felt like she was contributing to the war effort. She often said, "it was an exciting time and everyone thought their jobs were important." Thelma turned 97 at the time of the interview.

1945

Girls and a Milling Machine

Ethel Pausch, Mary Ellen Gossage and Marguerite Lodde, who worked in job evaluation, left the serenity of their typewriters and paperwork a couple of times to peek curiously through the door that separated the office from the factory. What they saw there was frightening. Gigantic machines were gnashing their vicious teeth nosily, chewing up pieces of metal and whirling, sawing and sliding in a perfectly terrifying manner. The three girls were required to learn to run machines at the training school in order to have a better understanding of their job. The trembling with which they entered the school's machine shop grew to panic proportions by the end of the first day. However, the machine shop was just like home to the three office workers by the time they'd reached the milling machine course. Miss Pausch deduced, "I imagine they make some granular, flour-like by-product of the engine." Miss Lodde opined, "The surface grinder is much prettier than any other machine. It makes lovely sparks." Complex shop terms now are sprinkled plentifully through the vocabularies of the three office workers, who just four weeks ago were alarmed at the sight of a machine. They even nicknamed themselves Odie, Idie and Involuna.

1944

Left to right: Marguerite Lodde, Ethel Pausch, and Mary Ellen Gossage

A Shopworker's Suggestion Pays Off

Fanita Markland thought of a gadget to catch the chips before they clogged the screens of the Baush spotter she operated. A suggestion was filled out and returned by Markland (top). Vernon Lundeen, a suggestion investigator, went to Markland's machine to look over the apparatus (bottom, left). Her idea was accepted by the suggestion committee. M. M. Mitet, her foreman, handed her the award plaque and $7 in war stamps (middle).

1944

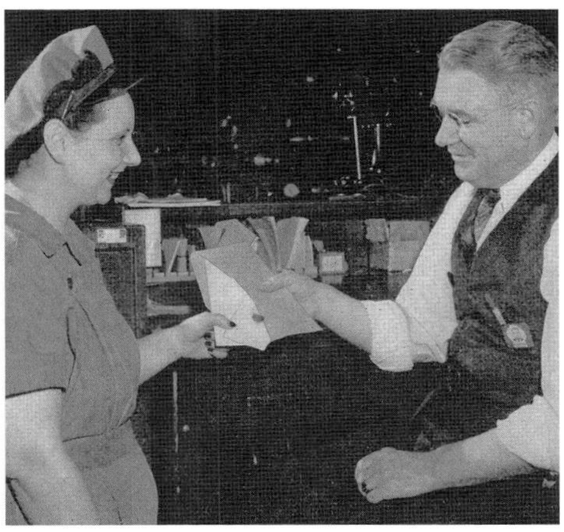

The Tool Crib

It was a library, but you won't find any Classics or Westerns here. The various department tool cribs were home to the thousands of tools that Pratt & Whitney employees required to produce engines. From gauges to grinders, an employee could check out just about any tool you could think of.

1944

Wilma Bock Doelling at Pratt & Whitney Kansas City

No person could hope to remember all 25,000 of the items kept in stock by the master crib and its 31 subsidiary branches, the departmental cribs on the factory floor. The variety of the stock made the average department store seem picayune. Because of the staggering number of items in the cribs, it was impractical to give the new stock keeper any preliminary training at the school. The new employee worked in the master crib, helping the old hands and familiarizing herself with the stock, until she learned the routine and was ready to be assigned to a specific job.

1944

Maxine Lawrence holding a setting ring gauge

Demonstrating Their Craft

A Pratt & Whitney Aircraft public exhibit, featuring a 2800-B engine and a cutaway 1830 engine loaned by East Hartford, gave all but a handful of the thousands their first chance to study a big aircraft engine. From the Pratt & Whitney school, William Grier selected (left to right) Tommy Anderson, Alfreda Seavy, Lucille Tagtmeyer and Juanita Ballard to demonstrate the steps in inspection.

1944

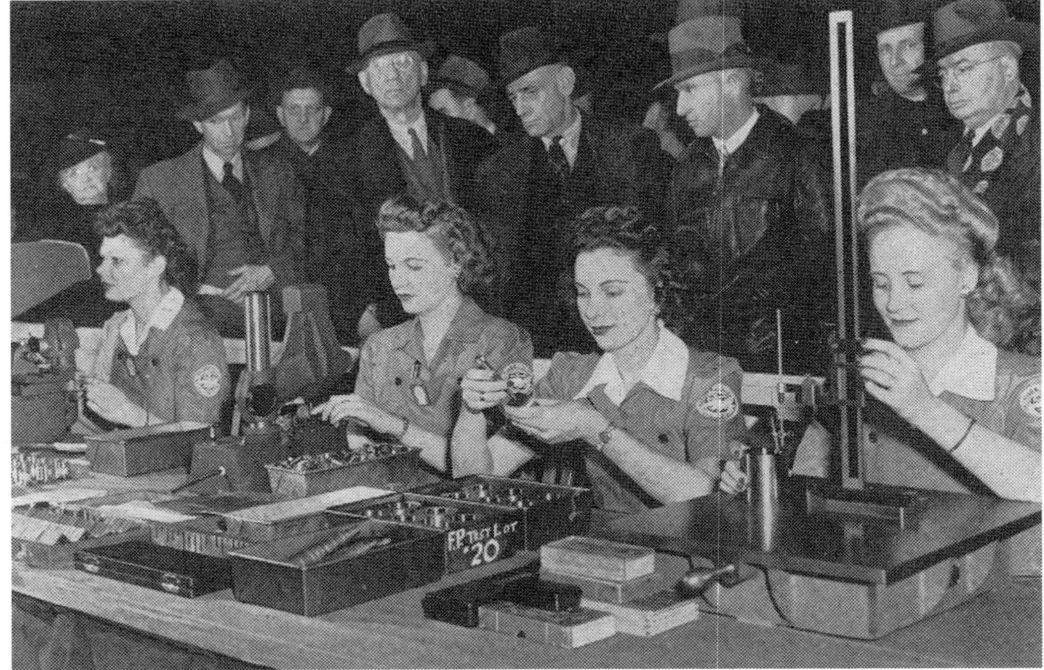

Norma Miller

After a full day of exacting work over a drafting board in the tool design section of production engineering, Norma Miller had a full evening ahead. A violinist in the Kansas City Philharmonic Orchestra, she was the only woman member who held a war job. It wasn't a great shift from pen and protractor to a violin. Whether working on a symphonic score or on a 2800-C engine, Miller pointed out, every part must fit and each must be perfect to fit.

NORMA MILLER

Research Aides Prove Valuable

Nowhere had the value of engineering aides been more apparent than in the Research division of United Aircraft Corporation, which had made use of female aeronautical engineers for several years and had long been convinced of their ability. The engineering aides, who had taken six-week courses at Connecticut College in New London, and the graduate women aeronautical engineers together comprise more than 40 percent of the division's technical personnel. The division's aides were engaged in all branches of analysis, research and engineering services, including both desk work and the operation of complex experimental equipment. Modifications of future or current aircraft caused by changes in the engine, propeller or operating requirements are analyzed by these aides. Although opportunities for permanent engineering posts would exist after the war, most of the aides intended to marry and settle down when peace came or enter other fields of endeavor. In the meantime, they were twisting knobs, watching dials, pouring mixtures, taking notes and consulting slide rules in a thousand jobs that needed to be done if progress was to continue, and a war was to be won.

ENGINEERING AIDES WORK ALONGSIDE ENGINEERS IN MANY JOBS. KATHRYN HALL CHECKS A LAYOUT.

1944

Engineering Aides: Women at Work on Important Jobs

Trim young women, some watching complicated testing machinery, others working with chemicals and Bunsen burners and still others bending over drawing boards or calculating tables with the ever-present slide rule in their hands, were disputing the old adage that "a woman's place is in the home" and at the same time making an unusually valuable contribution to the war effort.

This page, from top to bottom: Ann O'Rourke tested a carburetor scoop in the experimental test department at Pratt & Whitney division; Jane M. Grey was an illustrator for Pratt & Whitney's installation department; Caroline Glaskowsky checked a propeller governor.

Facing page, top to bottom: Josephine Dees worked in the material development laboratory at Pratt & Whitney division; Susan Baer drew spinner assemblies in the drafting room at Hamilton Standard Propellers Division; Marion Kingston and Barbara Ramsdell were making strain tests in the structural test laboratory at Chance Vought.

1944

Precisionists: Gauge the Gauges

Measuring devices were accurate to a millionth of an inch and were calibrated with size blocks that represented the absolute standard. The gauge standards department adjusted, repaired, and certified the accuracy of the gauges used by inspectors in all departments. Every gauge was checked regularly.

1944

Anna Bell Wolf and Wanda Zeiger

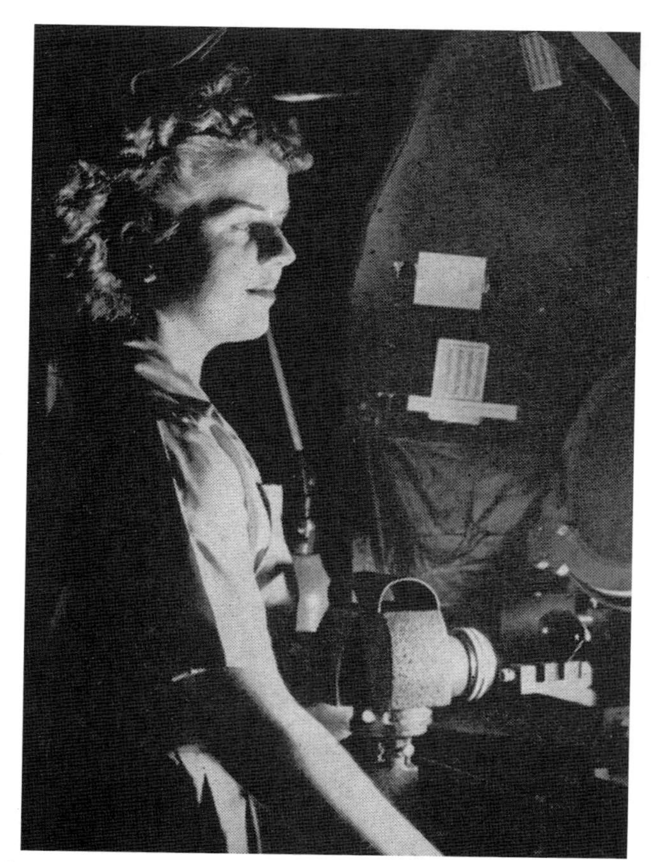

Virginia Russell

Edna Lonegan

Clarice Moon, Betty Cooper, and Dorothy Flaigle (unknown order)

Donna Jean Willcoxon, Emma Skidmore, La Verne Ferguson, and Mary Helen Parker (unknown order)

Eagle-Eyed Ladies of Inspection

Inspection of R-2800-C engine parts was not confined to process checking during manufacturing. The pieces underwent many other tests before they were finally approved for the last engine test. After the oil bath, the pins were examined for cracks by Edith Fauch and Evelyn Ray (left). Norma Watson (right), clerk, recorded the serial numbers of the parts.

1944

Atom Inspectors

The process of analyzing atoms via spectrographic film was a detailed undertaking. Marjorie King (top) put a negative through a quick drying process. Once the film's data had been tested, the results were taken to the calculating table by Janet Falkenberg (bottom, standing). The results of the inspection were computed, and then Lucyle Skinner (bottom, at the drafting board) made a graphic account from the calculations of the metallic content found in various samples of the part.

1944

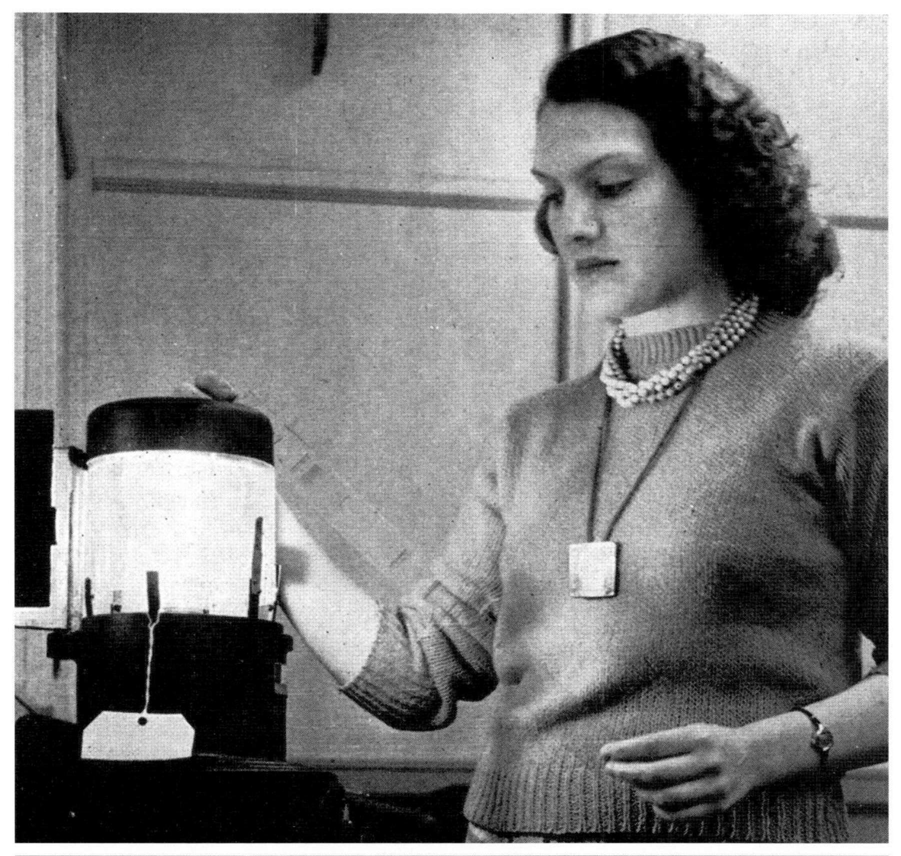

Naval Inspection

The women of Pratt & Whitney's Naval Inspection Service clerical department wearing navy blue and gold. Left to right, they were Anita Polson, Erna Hartmann, Geraldyne Davis, Gerry Wright and Margaret Lamprecht. Design changes made in the regulation Naval Inspection Service uniform by these women were being adopted by other Bureau of Aeronautics inspection offices. Factory inspectors wore the uniform without brass buttons and with open-throat shirts.

1945

Communications: On the Brink of a New Era

Communications was the nervous system of the Pratt & Whitney plant. Impulses of action that may have originated in Washington, or in the Pratt & Whitney Aircraft Missouri executive offices, or on the factory floor were received at the largest private switchboard in Missouri and transmitted to the 1,050 phones in the offices and factory. On the day shift, Lena Caraway and Marguerite Van Bibber (left to right) handled a major portion of the numerous calls routed through the board. Other operators were Florence Dugas, Mary Fritsinger, Flossie Deverell, Shirley Osborne, Dorothy Cummings, Catherine Trussel and Helen Vaughn. Mae Moske was the chief operator. Another important nerve center was the telegraph room. All telegrams, teletypes and direct wires to and from Hartford were received and sent by Peggy Barker, Orlean Garnett and Betty Johnson.

1945

Pratt and Whitney

1960, spring: First run of a prototype industrial gas turbine.

1961 (April): First run of the JT8D engine.

1961 (May): First run of the J58-P-4 engine for the SR-71.

1961 (May): First flight test of the Pratt & Whitney Canada PT6 engine.

1961 (October): First launch of the Rocketdyne-powered Saturn 1.

1964 (December): First flight test of the TF30, Pratt & Whitney's first afterburning turbofan. First flight of the J-58-powered SR-71.

1966 (March): First run of the JTF-17, Pratt & Whitney's supersonic transport engine.

1966 (December): First run of the JT9D engine, the first Pratt & Whitney high-bypass turbofan.

1967: First run of the JT15D.

1970 (February): JTF22 demonstrator chosen as basis for F-15 engine, the F100.

1970 (December): First flight of the F-14 Tomcat.

1972 (July): First flight of the F-15/F100.

1974 (February): First flight of the YF-16 prototype.

1974 (August): First run of the JT10D engine.

1977 (March): Flight test of the JT8D-200 series prototype.

1979 (October): MD-80 first flight.

1981 (December): First run of the PW2000.

1982 (February): Development flight testing initiated for the PW100 engine.

1983 (December): International Aero Engines (IAE) formed.

1984 (April): First run of the 94-inch-fan PW4000.

1984 (December): PW2000 enters service.

1985 (December): First run of the V2500 engine.

1987 (June): Entry-into-service (EIS) of the 94-inch-fan PW4000.

1989 (May): EIS of the IAE V2500.

1960s

1970s

1980s

Family Tree (Continued)

1990 (August): PW300 series engine certification.

1991 (April): Pratt-powered YF-22 prototype wins Advanced Tactical Fighter competition.

1991 (December): MP200 series certified.

1992 (December): First run of the F-119 engine.

1994 (December): Entry-into-service of the PW4168. EIS of the 100-inch-fan PW 4168.

1995 (May): PW4084 is the first engine certified for 180-minute ETOPS at EIS.

1995 (June): PW4084 engine entry-into-service.

1995 (December): PW500 series certified.

1996 (August): Pratt & Whitney and GE form the Engine Alliance to develop engines for super first flight of F-22/F119.

2000 (August): First run of the PW6000 engine.

2000 (September/October): First flight of Pratt & Whitney–powered Joint Strike Fighter (JSF) prototypes.

2001: First run of Advanced Technology Fan integrator engine, prototype for the Geared Turbofan™.

2001 (October): Pratt given formal contract for F135 engines intended for the JSF.

2002 (February): Pratt-led team wins Collier Trophy for JSF lift fan system.

2003 (June): Pratt & Whitney GDE-1 is the first flight-weight scramjet to attain Mach 4.5 in tests.

2004 (March): First run of the Engine Alliance GP7000.

2004 (December): First PW6000 flight on an A318 airplane.

2005 (November): Pratt & Whitney forms a new unit, Global Service Partners, to rapidly expand its service business.

2005 (December): Pratt & Whitney Canada certifies the initial PW600 for the emerging very-light jet market.

2006 (November): First flight of the GP7000/A380.

2006 (December): First flight of the F-35 JSF.

2007 (October): Mitsubishi selects the Geared Turbofan™ for a new regional jet.

2007 (November): First runs of the Geared Turbofan™ engine.

2008 (February): Bombardier offers a C-Series jet with a Geared Turbofan engine™.

Next Generation Begins

1990s

2000s

CHAPTER THREE

THE POST-WAR YEARS

The post-war years ushered in decisive changes in the workforce. Whereas in the 1940s, a woman's role in the factory was seen as a largely temporary phenomenon, a 1950s female employee had a decidedly different expectation for her career path. Pratt & Whitney identified the potential and dynamism of jet technology as well as the value of women employees. A professional woman, like the jet engine, had become and continues to be a vanguard in the success of Pratt & Whitney.

With the 1950s came the dawn of the Jet Age, when we said goodbye to an era of piston-powered commercial aircraft and said hello to the jet-powered era. Invented in the 1930s, applied militarily in the 1940s and made commercially viable in the 1950s, the jet engine's mark on aviation endures. Having proven itself in the era of piston power, Pratt & Whitney was determined to keep its position as the preeminent producer of aircraft powerplants. United Aircraft Chief Engineer Leonard Hobbs and his breakthrough axial flow J57 turbojet were instrumental in establishing Pratt & Whitney's reputation as the standard bearer for jet propulsion.

The subsequent decades saw technological advancements that propelled Pratt & Whitney to even greater heights. The company's transformation to a jet manufacturer in the 1950s, leveraging of computer power in the refinement of the J57 turbojet into the efficient and prolific JT3D turbofan in the 1960s, nimbly negotiating the energy crises of the 1970s, and searching for productivity efficiencies throughout the 1980s all contributed to Pratt & Whitney's continued success. Pratt & Whitney's embrace of emerging, next-generation technology mirrored the successful transition of women as the new workforce of technical professionals in the company, while they continued to promote educational advancement and charitable giving.

The Ladies of Payday

Nose to the grindstone also applies to the back-office ladies. Machinery is set into motion when workers turn in time cards. An IBM machine figures the work time while other machines compute the net salary. The sweetest sound in Pratt & Whitney Aircraft is the blast of a field whistle to let everyone know it is payday. Salary checks are sorted and bundled in the paymaster's office before delivery to employees in the shop. Heavy coin boxes roll into the shop as pay hour nears. A woman employee takes signed checks while a guard hands out cash.

1955

An IBM machine figures the work time.

Machines compute net salary.

Salary checks being sorted.

Field whistle blown on payday.

Heavy coin boxes roll into the shop.

Seeing Is Believing

Optical devices could be found in nearly every department of Pratt & Whitney Aircraft. With accuracy as the keynote, they helped the machinist to read hairline markings on Vernier scales and micrometers. Optics detected tool marks and infinitesimal flaws that might eventually lead to metal fractures. Optics were also used in materials controlled by using a Metallograph.

1955

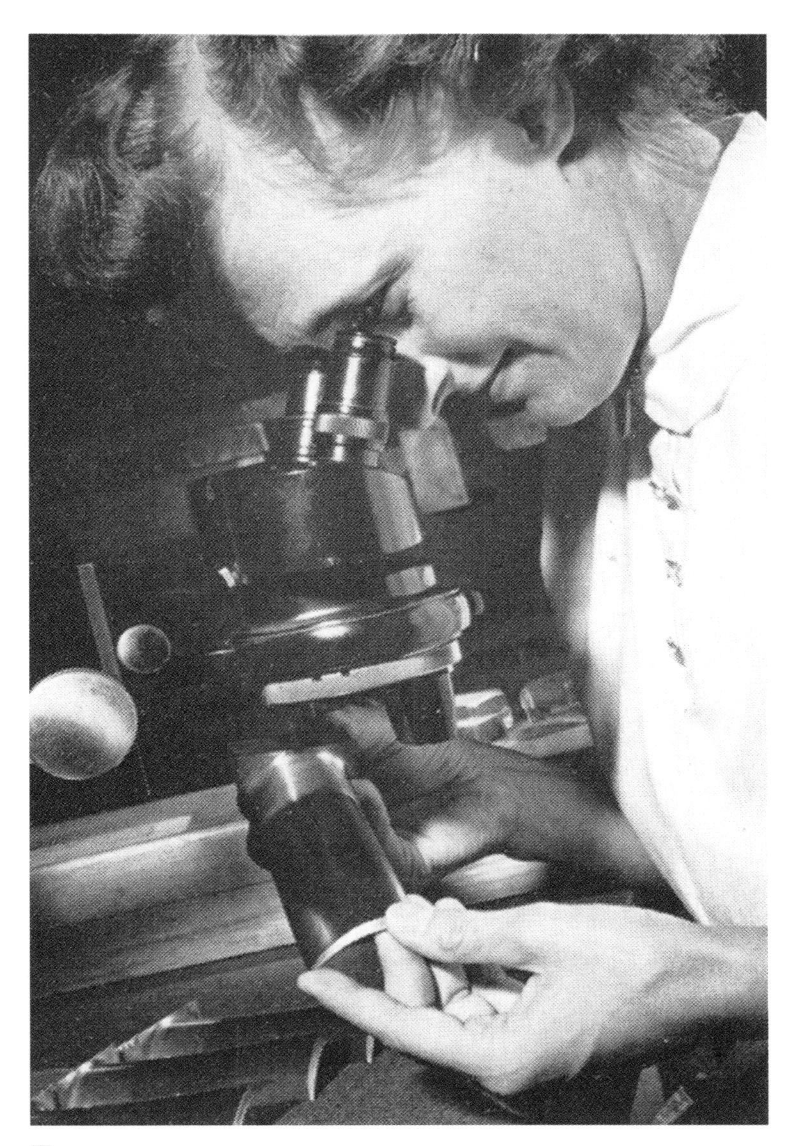

Turbine blade

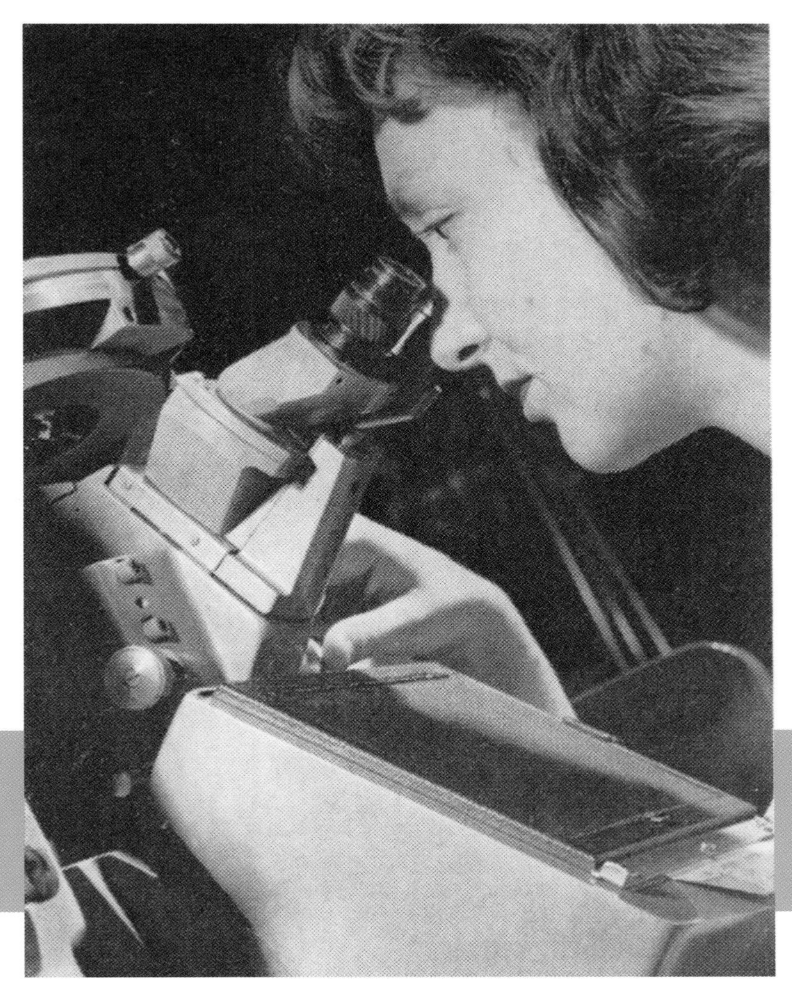

Metallograph

Scholarships: Building the Future

So quickly had the United Aircraft educational program been expanded that relatively few people were more than vaguely aware of its scope. The women pictured were among the September registrants at the Rensselaer Polytechnic Institute Graduate Center in South Windsor. At the time, 24 scholarship holders were in college, and the next year it was expected there would be 35 in force. In June 1956, the undergraduate program was broadened to include an annual scholarship for a daughter of an employee of any United Aircraft Corporation division.

1956

Data Processing Professionals

Vital information could be stored and transferred through the use of punch cards. Data could also be relayed directly between the design, experimental test and Willgoos Laboratory engineering offices at Pratt & Whitney Aircraft and the computation laboratory. At top speed, these transceivers (transmitting and receiving) could pass coded information back and forth through regular telephone lines at the rate of 800 characters a minute. Magnetic tape recorded information, which was available for immediate processing. Punch cards were fed into a card reader to transfer data to the memory units.

1957

Magnetic tape

Punch cards

Women Master Technology

The woman on the right used a flash butt welding machine, which melted edges of parts and then forced them together, producing a strong bond.

1956

Eileen Milnerstadt carefully inserted coils into a stator.

1958

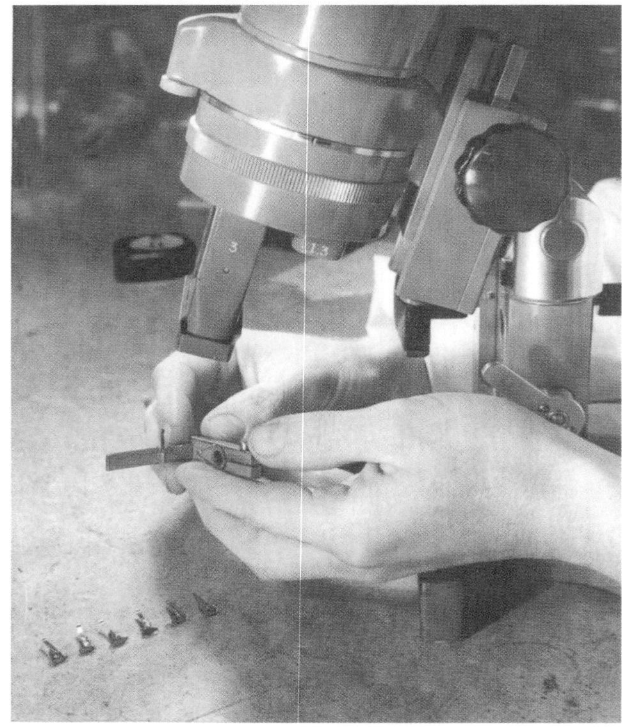

The assembly of potentiometers at Ketay department called for fine precision work.

Collaboration Around Computers

Computer languages were relatively new in 1960. Because of this, they were constantly being improved upon. Wherever there was a sizable computer installation, people were studying and experimenting, attempting to find better ways to talk to machines. The data processing center at United Aircraft Corporation's Research Laboratories had two IBM 704s and a Philco solid state 2000.

The center employed seven people dedicated to program research. New ideas for computer programs came from specialists such as Kathleen Gormley and John Hetherington of United Aircraft's program research group, seen here at a Philco 2000.

Kathleen Gormley and John Hetherington using a Philco 2000

1960

Another early computer user

Charity at Work

From giving blood to giving a portion of their paycheck, the philanthropic spirit of Pratt & Whitney employees was unparalleled. More than 3,800 pints of blood were donated in 1964 by employees at Pratt & Whitney's main plant. The spirit of giving also extended into the civic arena, with more than 1,400 employees soliciting funds on behalf of Community Chest, Red Cross and United Fund organizations. With a contribution goal of $2,500,000, every volunteer was as motivated as ever to make the year's pledge drive a success. Historically, as the records demonstrated, women volunteers were at the forefront, establishing Pratt & Whitney's enviable culture of giving. The 1964 archives proved it to be a banner year for women as charitable champions, noting their involvement from every department and location.

1964

From left to right: Shirley Dallman, personnel investigation, Southington; Joanne Branigan, materials accounting; Danielle Perras, production engineering, North Haven; and engineer Kathleen Coleman, materials accounting

From left to right: Linda L'Heureux, personnel investigation, Southington; Joyce Sirois, spare parts accounting, Podunk; Janice Weglowski, master mechanic's group, East Hartford; Betty Parent, Apollo Project, Orchard Building; and Dorothy Fargo, engineering illustration, East Hartford

Building the Mentoring Tradition

The Junior Achievement youth program was composed of high school students who organized and managed their own small-scale businesses under the guidance of adult advisors from local businesses and industries. Local young women, seen here, benefited from the wisdom of Pratt & Whitney employees. These photos represent an early example of future programs that would promote women in aerospace careers.

1964

Anne Gabunas

Eileen Kelleher

Adventure Awaits

Mary Sawers, who worked for 20 years at United Aircraft, left for Afghanistan on a two-year teaching assignment for the Peace Corps. Sawers, who worked in the office of Perry W. Pratt, UAC vice president and chief scientist, said before leaving: "I am enthusiastic about the Peace Corps job. I took a three-month training course after my retirement last January, and I am eager to get on with the new work in Kabul. I will teach general business subjects to students of high school age."

1965

Mary Sawers

Quality in Craftsmanship

Sterile, laboratory-like surroundings were vigilantly monitored by technicians and ensured consistent results over the array of tools tested. The high degree of accuracy was necessary because the development of lighter, more powerful and sophisticated engines required a greater precision fit between parts. Louise A. White, for example, used optical flats; the instrument gave a highly accurate measurement of flatness. 1965

Lydia Casson

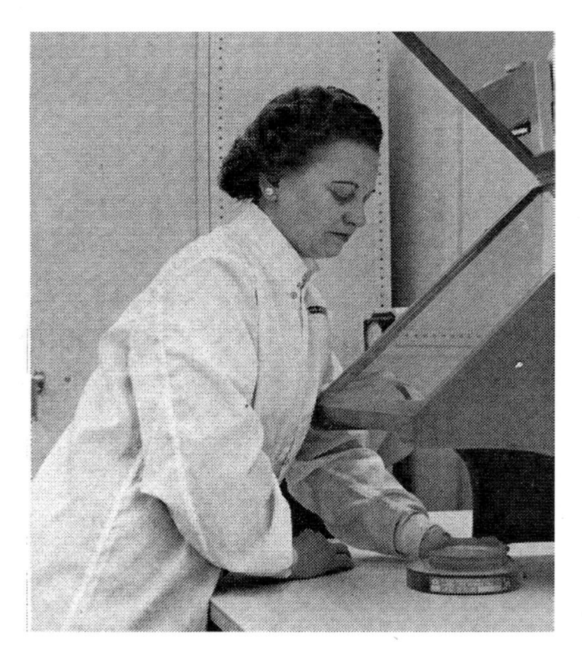

Louise A. White

Third Shift

Material had to move around the clock at the North Haven plant. Marian Robinson, an Aircrafter one year, was the sole typist on the Dispatching Department's third shift. 1966

Marian Robinson

The New Learning Curve

Loretta Skold embraced the challenges of being an instructor at the Training School. Did she run into problems because she is a woman? "It might be an advantage," she said. "Perhaps it is because men are used to being taught by women in their grammar school and high school days. I explain the rules when the course begins, and the students usually abide by them." Are men easier to teach than women? Yes, Mrs. Skold said. "Women need more encouragement and are more afraid of looking foolish if they make a mistake. On the other hand," she said, "women adapt more quickly and are less clumsy."

Here, Mrs. Skold showed Robert Silvia, D978/518, how to complete a layout inspection of an accessory housing cover. Silvia was in his last week of an eight-week course at the Machine Training School when the photo was taken.

1966

A Pratt & Whitney Milestone

The 5,000th Pratt & Whitney Aircraft JT3D, the world's most widely used turbofan engine, left East Hartford in 1967. Evolved from the J57 turbojet that ushered in the commercial jet age in October 1958, the JT3D engine was the standard powerplant for various versions of the Douglas DC-8 and Boeing 707 and 720 jetliners. Under its military designation, TF33, the powerplant was found in the Boeing V-52H Stratofortress and Lockheed C-141 StarLifter.

The JT3D was certificated by the Federal Aviation Agency in 1960 and entered regular commercial service in March 1961. At the time when these photos were taken, it, and the military version, had accumulated more than 23 million flight hours. More than 60 airlines had ordered over 800 planes powered by the JT3D. It developed up to 21,000 pounds of thrust in its military version.

When American Airlines put its JT3Ds into commercial service, the time between overhaul (TBO) was 800 hours. By January 1967, the TBO had been upped to 10,000 hours—the highest TBO given to any aircraft engine in the world. As engines continued their path of ever-increasing feats of performance, women were still transitioning from the pin-ups of the 1940s to the recognized professional in 1967.

1967

The Recognition Story Is Still Evolving

Working Girls

There is nothing quite like a girl. She makes even the summer day seem brighter; helps the work day pass more pleasantly. Her contagious smile is pleasing to those around her. When she tosses her head or laughs, she is a joy to watch. She inspires poets and photographers in all the year's seasons, particularly in summer. Luckily, there are lots of pretty girls working at Pratt & Whitney Aircraft.

1967

Ethel Ludlow

Kay Pavliscsak

Patricia Phelan

Growing Professionals: Passion for Precision

Winifred Shea, right, and Eleanor Stender, left, were two of four women draftsmen at the Middletown plant.

Statistical quality control technician Marge Bennett on the second shift in Area 23 at the North Haven plant.

A former Pratt & Whitney Aircraft scholarship winner, Donna Nedelka was a part of program analysis in Advanced Power Systems. Her father, Donald Pratt, was a leadman in D-34.

1967

Testing of Technology

Orlaine Hartman worked with stress and flow analysis as well as vibration and fatigue testing.

Barbara Gledich tested a fuel pump for a Pratt & Whitney Aircraft jet engine.

Betty Ann Sayles was an assistant tester at the Willgoos laboratory.

1967

A Place for Everyone

The year 1967 was a pivotal one for women. The climate of change proliferated into every woman's life through historic legislation that articulated the rules of workplace equality. Fostered by the recently formed National Organization of Women, a "Bill of Rights for Women" was established, eventually becoming part of the renowned Equal Rights Amendment (ERA).

The ERA was far-reaching, including topics on sex discrimination and segregation in employment ads. By the end of 1967, four New York City newspapers, including The New York Times, set a new precedent by de-sexigrating their "help wanted" ads. The era of professional equality arose . . . with Pratt & Whitney's early adoption of the inclusive work environment, already ahead of the social climate.

1967

A New Adventure

Linda Barton, Eunice Monroe, Lucile Pernell and Paula Ruffin were the first women to enter an apprentice training program in Pratt & Whitney Aircraft history. The training program at Pratt & Whitney Aircraft's training school at the Middletown plant was part of the three-year machinist apprenticeship, which involved instruction in basic machine familiarization, plus classes in mathematics, blueprint reading, machine theory and measuring tool instruction. Working with the men in the program had not been a problem for the girls. "The guys are really helpful," Eunice remarked, "but I'm glad there are three girls in this class instead of just me." Lucile said that being one of the first females in an apprentice program was a thrill, "as long as it goes the way it has been going." Kingsley Carpenter, supervisor of mechanical training, said, "The females are sparking the whole thing up, it's only tradition that had made the program strictly for men."

1972

Lucile Pernell measured a ring on an internal grinder. Looking on is Instructor George Olsen.

At the training shop, Eunice Monroe leveled a part in preparation for surface grinding.

Creating a Career

Thirty-year employee Muriel Stipek originally hoped to be a medical X-ray technician, but came to Pratt & Whitney Aircraft during wartime to work as a toolmaker. At the end of the war, Stipek left the company and went to Hartford Hospital, where she trained as an X-ray technician. In 1946, Stipek applied for a job in Pratt & Whitney Aircraft's medical center. There were no openings there, but the company was looking for technicians in production X-ray. She applied and was rehired. Stipek used various X-ray machines to inspect such parts as castings and fuel lines for defects. Stipek now found herself supervising 21 men. Her boss appraised her management style as "very businesslike, firm but fair."

1973

MURIEL STIPEK

EUNICE MARY GETCHELL

Quarter Century Club

Eunice Mary Getchell (Decker) moved to East Hartford, Connecticut, and started working for Pratt & Whitney in 1951, when she was 39 years old. She started on the assembly line, then worked as a timekeeper, and ended her career working in the tool crib. This is the first documented photo of a woman receiving her award for being recognized for 25 years of service. Her clock # was 81104.

1976

Purchasing Power

For the first time in Pratt & Whitney Aircraft history, five young women, all recent college graduates, had entered the purchasing department's buyer trainee program. As buyer trainees, the five women, Sherilyn Morris, Susan Larose, Diane Hinchey, Susan Read and Patricia Reagan, performed a variety of purchasing assignments under supervision in the experimental and non-product materials sections. Susan Read had been a stenographer at Pratt & Whitney Aircraft. "I wanted more of a challenge," she said. Susan went to college where she earned a degree, which enabled her to enter the training program. Some of the trainees had family ties with Pratt & Whitney Aircraft and thought that they would see if there might be any interesting job openings. "You must establish your role as a buyer, not as just a woman," Sherilyn Morris commented. After these five new buyer trainees had been on the job for a while, reactions toward them were beginning to change. "Everyone is taking a different attitude toward us," said Sherilyn. Upon successful completion of the training program, the women would be buyers, handling order requests, securing prices from vendors for the requested merchandise, and analyzing the effectiveness of the purchase. They would follow up all purchases to ensure delivery and quality of merchandise.

Locating a part on a blueprint are, from left, Patricia Reagan, Diane Hinchey and Susan Read.

1973

Women in Engineering

Teresa Hinds of engine design in North Haven supervised three sections of the department including reproduction services, drafting support and clerk typists.

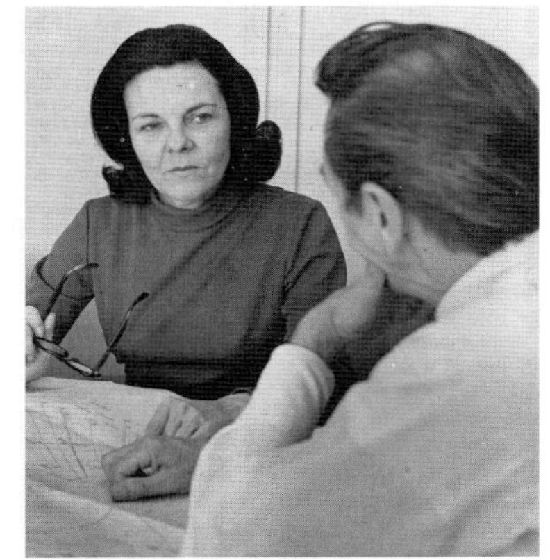

An employee of Pratt & Whitney Aircraft for 30 years, Juliet Coyle prepared the reports required by NASA, the Air Force, and the Navy for engineering development contracts. As Supervisor of the technical report group in engineering, she also prepared presentations for technical meetings.

A senior materials engineer at MERL inserted a solution cell into an X-ray machine for analysis of alloying components.

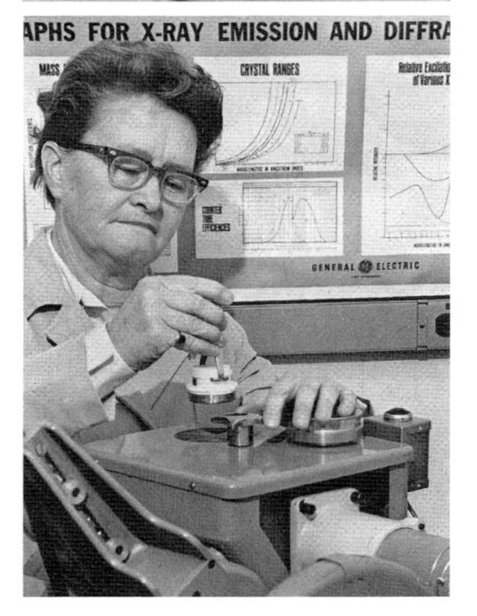

Charlotta Simpson, who in 1973 had been with Pratt & Whitney Aircraft for over six years, worked as a materials engineer in the Materials Engineering Research Laboratory in East Hartford. Here, Charlotta analyzed alloys for trace metals.

1973

Stealth Talent

It's a good thing awards and honors are presented publicly; otherwise the family, friends and work associates of certain self-effacing honorees would never find out. Corienne B. Stevens was one of 32 women in America to be singled out for outstanding achievement in business and industry by the YWCA in June 1977. "I'm not a 'women's libber' as such," she said. "I believe, however, that women have excellent minds just as men do. As far as I'm concerned, if you use your intelligence, you can do just about anything you want to do." Stevens was a metallurgist whose work, since joining Pratt & Whitney Aircraft's Government Products Division in Florida in 1961, had centered on the study of nickel-based superalloys. Her research contributed to the development of the F100 jet engine. Stevens directed all chemical, crystallographic and metallographic analyses at Government Products. She was also responsible for research supporting the development of advanced jet and rocket power plants.

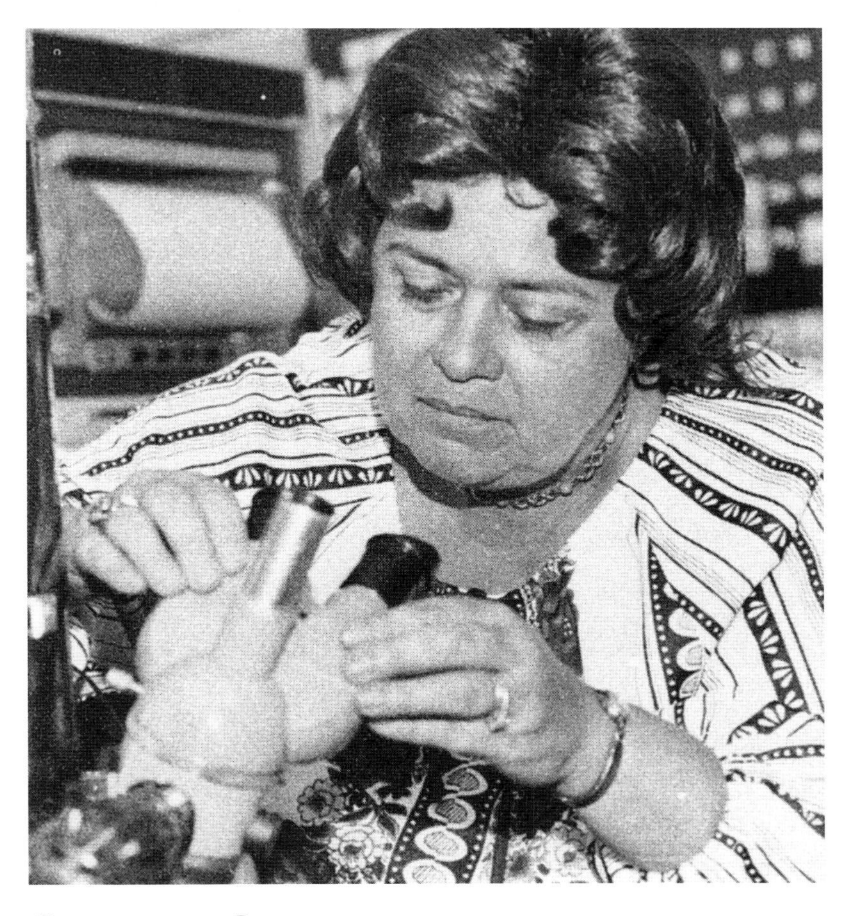

CORIENNE STEVENS

Stevens learned in early May that she had been chosen to be honored—but she kept very quiet about it. No one learned anything from her. Even her husband didn't find out until two weeks before they were to leave for the presentation dinner at the Waldorf-Astoria Hotel in New York City.

1978

Emerging Talent

Forty-five college scholarship awards to children of employees were announced in 1979 by United Technologies. The total awards to the 1979 winners over the next four years may have reached $360,000. Each scholarship was up to $2,000 tuition and academic fees for each of four years' full-time study leading to a bachelor's degree at an accredited college or university. Pictured are women scholarship winners who are children of Pratt & Whitney Aircraft Group, Manufacturing Division, Commercial Products Division or Research Center employees.

Top: Jennifer Beckett, Andrea Davis, Karen Davis, Michele E. Norton
Bottom: Aria Harrison, Cindy Johnson, Suzanne Lafreniere, Mary Michele Lake

1979

Into the 1980s with Confidence

Guests being shown materials at the North Berwick facility

Connecticut Employee Named a Suggester of the Year

Rhonda Pease, a Manufacturing Division, Middletown, Connecticut, plant inspector, won the Pratt & Whitney "Suggester of the Year" award for 1982. Pease saved the company more than $450,000 by suggesting a change in an inspection procedure, which simplified the filling out of high volumes of paperwork. "I try to make things on the job easier for me and other inspectors. If we keep the company going, we'll keep our jobs going, too," she said. A total of $5 million in savings to the company and $765,000 in awards to Connecticut and Maine employees resulted from voluntary participation in the cost reduction program.

1982

Charity Is Still a Part of Who We Are

Joe Kozlin, left, the Government Products Division (GPD) employee who organized the Gold Coast Marathon for the benefit of Special Olympics, presented a trophy to Norma Blackwood of Miami for being the first woman to cross the finish line.

Runners crossed the Blue Heron bridge in Singer Island, Florida, on their way to the finish of the 13.1-mile race.

1982

When Rebecca Webb, North Haven bench assembly operator, dreamed, she dreamed big. The 16-year Pratt & Whitney employee and former youth clubhouse director launched a drive to raise funds for a $3.2-million New Haven, Connecticut, youth center complete with roller-skating rink, 450-seat theater, stores, job center and restaurant, a project grown from the humble beginnings of a youth clubhouse she established in her New Haven garage in 1979.

Rebecca Webb

1983

Sky's the Limit

Laurie Thompson was married, a mother and an administrative assistant in Pratt & Whitney's Human Resources Department at East Hartford. Soon, she would finish earning her bachelor's degree from Eastern Connecticut State University in business administration, concluding a 13-year education adventure she began at Manchester (Conn.) Community College. Despite the hardships of combining a full-time job, family and school, Thompson was unequivocal: "The benefits override the pain."

A 16-year veteran of Pratt & Whitney, Thompson took advantage of UTC's Employee Scholar Program. The paid-time-off provision really helped make evening classes less stressful. "I can leave the office an hour and a half early to give me that time to make an early class," she said. "I'm not in school until 10 o'clock at night like I was before."

The Employee Scholar Program's streamlined tuition payment system also helped Thompson stay focused. Tuition and book expenses of up to $1000 a semester were paid directly to Eastern Connecticut by Pratt & Whitney.

Many women, like Thompson, benefited from the Employee Scholar Program while still being able to manage their households. By the late 1990s, Pratt & Whitney had helped women broaden their professional aspirations, making the company a preferred employer of choice for future aero-manufacturing experts and innovation pioneers.

LAURIE THOMPSON

1996

PRATT & WHITNEY

CHAPTER FOUR

THE TAKE-OFF

The courage, grit and grace demonstrated by Jacqueline Cochrane throughout the early years of aerospace endeavors is ever-present in the women of Pratt & Whitney today. More groundbreaking professional firsts occurred for Pratt & Whitney's women within the last 20 years of Pratt & Whitney's more than 90-year history than at any other time. As women began to increasingly populate the professional world, they found that, particularly in the aerospace industry, success was bound only by their imagination and hard work.

Pratt & Whitney's ability to adapt to changing contemporary workplace themes helped provide women work choices that promoted their permanent place in the workforce, forever eliminating the concept of women as temporary staff. The 1996 article that closed the previous chapter about Laurie Thompson earning her bachelor's degree as part of the UTC Employee Scholar Program is an example of the dedication exhibited by Pratt & Whitney and the women who advanced their careers. Women's choices made in juggling work and home life were self-evident, and the hard work has resulted in professional achievement at the highest levels.

Technology frontiers were now being infused with newly credentialed, technically capable women engineers, subject matter experts, and manufacturing specialists. The era of technical equality had begun.

Pratt & Whitney is again making history by transforming aviation through technology and innovation across its engine portfolio-including in the commercial market with its game-changing Purepower® Geared Turbofan ™ engine, and on the military side as evidenced by its F135 propulsion system on the F-35 Lightning II. Like World War II, when Wasp engines needed to be manufactured at unprecedented levels, today's manufacturing demand is as if history is repeating itself. Only this time, women are ready and positioned to answer the call.

The spirit of the pioneer is just as strong in today's work environment as it was in the earlier days of flight firsts. On the next pages are brief samplings of contributions by women employees at Pratt & Whitney, who are recognized for breakthrough impact, innovation, technical prowess, mentorship and years of impeccable service while they contribute to the company's future runway for success.

Pratt & Whitney Fellows

The Pratt & Whitney Fellows Program was established in 1995 to recognize and reward the company's technical experts and to enhance technical excellence. A Fellow is a designation of distinction, whereby appointments are driven by business and technical need. Fellows play a key role with defined responsibilities in identifying and closing technical gaps, developing technical bench strength and solving today's most critical problems.

The Fellows Program continues in its mission to expand professional development, foster technical excellence and promote leadership opportunities within Pratt & Whitney.

Attributes of a Fellow

The Pratt & Whitney Fellows program recognizes individuals who have demonstrated exceptional knowledge, judgment and competence in their technical discipline. They are recognized not only by Pratt & Whitney, but also their industry peers as valuable contributors and experts in their field.

Mentors

As mentors, teachers and leaders, Fellows directly affect the technical excellence of Pratt & Whitney's products, processes and people.

Outstanding Technical Achievers

A commitment to technical excellence is a distinguishing characteristics of Fellows, along with significant service to their company and the industry as a whole.

Process and Discipline Experts

The Fellows program allows designated individuals to enjoy the benefits and privileges befitting their considerable contributions to the company while enabling them to define their career path to continue their professional growth and develop further in a specialized technical discipline.

Industry and Professional Society Leaders

Pratt & Whitney Fellows have the opportunity to share their expertise outside the company through collaboration with government, academic and corporate entities. Fellows serve on industry committees and consortiums, participate in government and university research, and assume the lead in presenting technical papers and participating in technical conferences.

First Female Associate Fellow: 1996

Mary Austin was named an Associate Fellow in the second year of the Fellows program, recognizing her significant contributions over the previous 19 years to business operations in the areas of advanced controls and intelligent processing. This achievement acknowledged Austin's comprehensive understanding of the company's products and her ability to apply her knowledge to the designing and building of intelligent systems. She was a major contributor to the design and development of a Fuzzy Logic Adaptive Control Welding System, which, at the time, was the most advanced welding system in the United States.

MARY AUSTIN

Before being recognized as a Pratt & Whitney Associate Fellow, Austin was the recipient of a National Science Foundation Industry Representative Award. Her contributions as a patent owner and author of technical papers further exemplified her knowledge and expertise and showed her commitment to and leadership within the industry as a whole.

SUSAN BROWN

First Female Fellow: 1997

One year later, Susan Brown was recognized as the first female Pratt & Whitney Fellow. During her previous 20 years with the company, she made significant contributions in the areas of lubricants and lubrication systems in the Engineering division. Among other accomplishments and advances, she contributed to the success of the F119 engine program.

According to her award citation at the time, on the development side, Brown provided a Pressure Differential Scanning Calorimetry Method for determining auto-ignition temperatures of engine oils, and also completed significant research into the factors associated with the formation of coke deposits. She also introduced the concept of a single-fluid jet engine where one fluid can serve as both the fuel and lubricant.

First Chief Engineer: 2000

Lynn Gambill's Pratt & Whitney experience began in June 1979 when the company was recruiting at the University of South Florida, where she received her bachelor's degree in chemical engineering. At the beginning, Gambill was one of about 30 women engineers at Pratt & Whitney's Florida facility. "I was determined to do the very best job that I could, regardless of the assignment, and to always do more than what was asked for," she said. Gambill was hired to work in the structures group and supported engine lifting. She supported J58, J52, TF30, F100, FL10 and SSME turbopump production engines and was responsible for various types of structural analysis.

Lynn Gambill
2015

While in the structures group, Gambill became a project engineer on the Space Shuttle Main Engine program, and also worked on the high-pressure turbopumps (HPOTP and the HPFTP) as a Design Verification System (DVS) project manager. She then moved to the PW5000 program and was one of four first-design integration managers at Pratt & Whitney, as well as the first woman configuration control board (CCB) chairman.

In 2000, Gambill surpassed a professional plateau when she became the first woman to be a program chief engineer for the F119 program and, ultimately, achieved the job title of Operational Military Engines chief engineer.

She then transitioned into the support of all commercial engines, including the PW2000/F117, PW4000, V2500, PW6000 and GP6000 engines, in addition to the military engine models. Support of Next Generation Product Family (NGPF) also became part of her role as

Left to right: Stan Stevens, Randy Lamar, Chris Flynn, Bennett Croswell, Lynn Gambill, Jorge Alcorta, Tom Farmer, Gen. (Ret) Willam Begert, Howard Lindler, Gerry Rambo.

it continued through development. At this time, Gambill was involved in operations manufacturing as well as field support for all Pratt & Whitney engine models, both commercial and military engine products.

During the late 1970s and early 1980s, Pratt & Whitney grew its women's engineering presence. "In the company, there was one woman who had attained a middle management level in the materials and science organization of Engineering," Gambill said. "Since then, the company has provided management training and many career opportunities for women with technical capabilities." Women have been provided many responsible roles, including: flight test engineers; analytical performance engineers; engineering component integrated project team and project managers; structures, design and analytical roles; material and science roles; and roles in new technologies, such as additive manufacturing.

Today, there are female executives throughout the company, including in Engineering and Operations. "I can't imagine working on anything more interesting, challenging, and exciting than aerospace engines," said Gambill, an extraordinary pioneer, "and I appreciate this opportunity every day."

Award Winning Women

Pratt & Whitney's women have fostered a pioneering legacy through many avenues of recognition. As barrier breakers, many prestigious awards have highlighted the technological achievements of the innovators, role models and leaders in science, technology, engineering and math (STEM).

Award nominees are recognized in the company's innovation circles for the essential contributions they have made in developing new products and services, advancing technologies, striving for excellence and serving as mentors for generations of women to follow. Pratt & Whitney women engineers have an early and long history of winning awards through esteemed professional societies and institutions, including the Society of Women Engineers and the American Society of Mechanical Engineers, to name just a few.

The number of internal company and external industry awards given to Pratt & Whitney women for their achievements has grown tremendously over the years, so much so, that there simply isn't room to list them all.

Included on the following pages are two highly regarded and more recently added awards, showcasing Pratt & Whitney women's increasing recognition both within the United Technologies Corporation's family and throughout industry.

These awards have been highlighted for their holistic view of achievement as recognition opportunities that have expanded to include both leadership and community involvement for technical innovators awards and awards for manufacturing innovation and excellence.

Employee working titles in this section reflect the award recipient's position at the time of their award nomination.

Sponsored locally by the Connecticut Technology Council as the "catalyst for innovation and growth," the Women of Innovation® Award honors women accomplished in STEM fields who are involved in their community.

The Large Business Innovation & Leadership Award honors women who have managed a program, project or business unit in an exemplary way for a large corporation (greater than 500 employees) or a professional services firm. The honoree must also have shown devotion of their own time and resources – or marshaled those of their company or organization to support community organizations or nonprofits.

The Community Innovation and Leadership Award recognizes one outstanding woman each year who has created or encouraged a culture of giving at her company or who has volunteered her time outside the workplace.

Julia Hutchinson

Julia Hutchinson motivated a team largely composed of new engineers who implemented process improvements that resulted in jet-fuel savings of $2 million and the reduction of 50 tons of CO_2 in 2006 alone. Her technical contributions continued to make impact while she also initiated the first Women's Council at the Middletown, Connecticut, Engine Center with the goal of empowering women by helping them find resources to develop their careers.

JULIA HUTCHINSON
2009

AGNES KLUCHA
2011

Agnes Klucha

Agnes Klucha was the company's first Engineering Innovation Center program manager, an advocate for advanced manufacturing technologies. She formed the Rapid Additive Manufacturing Production Technologies Technical Council and led the integrated Advanced Fabrication System, linking Advanced Manufacturing Cells with Integrated Computational Methods Engineering.

Laura Holmes

Laura Holmes was responsible for growing world-class training and conference facilities in East Hartford, Connecticut, and Beijing, China, while being the first woman to be general manager of Pratt & Whitney's Customer Training Center. She led development of new Leadership and Maintenance Repair & Overhaul training programs for both Pratt & Whitney employees and the company's airframe customers, as well as development and implementation of eLearning technologies and learning analytics. Holmes also was dedicated to a number of community volunteer efforts throughout the Greater Hartford area.

LAURA HOLMES
2011

SHARON KILLIAN
2011

Sharon Killian

Sharon Killian, a shining example of a civil servant, researched town records to search for homebound seniors and needy families to whom she could provide gifts. She met with homeless men and women who lived under bridges and acted as a personal mentor, working to get them moved into protective shelters.

When the Salvation Army Drop-Off Center closed, Killian became a "one-woman drop-off center," delivering critical items to shelters. She initiated a coat drive that delivered 900 coats to meet local needs during the winter months and, in 2007, devoted the year to helping U.S. veterans by initiating a program to help them develop better job skills to find jobs.

Terrie Riggs

Terrie Riggs, in 2010, took on the role of aligning Pratt & Whitney to support Athena's Warriors, the first "all girl" FIRST Robotics team, recruiting mentors, hosting fundraising events and coordinating meetings. Riggs found a home for the team at Central Connecticut State University (CCSU), where engineering students mentor the team. Athena's Warriors were awarded the Rookie Inspiration Award.

As the Community Involvement Lead for the company's Women's Council, Riggs had become the face of the company in much of the community. She founded Women in Aviation, Connecticut Chapter, and was nominated to CCSU's Advisory Board Committee of Technology/Engineering Education.

TERRIE RIGGS
2011

Madeline Sola

Madeline Sola, a Pratt & Whitney structural engineer, came to the company as part of the INROADS program, designed to advance inner-city youth. As an active member of the Hartford chapter of the FIRST Robotics Team, Sola has enriched the lives of Connecticut's youth. Sola, a Hartford High School and Worcester Polytechnic graduate of modest background, is a real-life STEM role model, especially for inner city female students and an outstanding example of how hard work, learning and community service can lead to great things.

Kim McFadden

Kim McFadden's award was the result of her tireless efforts in mentoring, coaching and committing countless hours to Athena's Warriors, a young women's robotics team. She volunteered an enormous number of personal hours for build sessions competitions at the Hartford Convention Center and in Philadelphia. Contributing time during the summer months, McFadden supported an off-season robotics design project presented to CT FIRST, a community-based program focused on building a better world for tomorrow by engaging high school students in STEM. Her innovation in design engineering was critical to the team in designing and assembling an elevating arm mechanism.

MADELINE SOLA
2011

KIM MCFADDEN
2011

Tracy Propheter

Tracy Propheter, a senior designer for one of the most critical areas of Pratt & Whitney's game-changing engine has been prolific in her turbine product and process development. Submitting 62 invention disclosures, 35 are U.S. patent applications with seven fully issued. Patents of this nature strengthen Pratt & Whitney's competitive advantage. Her leadership also extends to involvement with the Society of Women Engineers. In recognition of her innovation and leadership, she was named Innovator of the Quarter by Pratt & Whitney's "Hot Section" Engineering Department.

TRACY PROPHETER
2012

DIONNE HENRY
2012

Dionne Henry

Dionne Henry led the Connecticut Pre-Engineering Program (CPEP) Board of Directors for three years, helping to serve as a catalyst to significantly change students' knowledge, attitudes and behaviors relating to the pursuit of STEM careers. Under her leadership, CPEP has added new programs and been recognized as a leader in minority youth educational program development. During her tenure, CPEP was awarded the Harvard Business School CT Alumni Chapter's Turbo Award. Henry actively fosters Pratt & Whitney employee involvement in CPEP and supports the National Society of Black Engineers.

Monika Kinstler
2012

Monika Kinstler

Monika Kinstler, a Pratt & Whitney principal engineer, holds multiple U.S. patents. Among her most important contributions to the company is the implementation of the innovative Materials and Process Engineering (MPE) Search Engine, which has the ability to "crawl" through tens of thousands of documents in seconds and for which she received a Pratt & Whitney Special Award. Her initiation of a Phase 2 of the MPE Search Engine increased document database capacity, refined and fool-proofed search queries and, most importantly, created a cultural shift to encourage archiving of current and historical documents, protecting and preserving the company's informational future.

Stacy Malecki

Stacy Malecki, when nominated, was the design integration manager for Pratt & Whitney, responsible for Configuration Management and Control of all company products and processes. Her 27-year career developed from its roots in Mechanical Design and Project Engineering with a bachelor's degree in mechanical engineering from Massachusetts Institute of Technology. Expanding her education with a master's degree in mechanical engineering and an MBA from Rensselaer Polytechnic Institute and assuming a variety of technical and leadership roles, Malecki achieved the distinguished role of Pratt & Whitney Fellow in Turbine Design.

Stacy Malecki
2013

Lynn Fraga

Lynn Fraga, as general manager in Customer Service Operations for Commercial Engines, used her engineering and project management skills to lead Customer Service infrastructure upgrades in their readiness for the Next Generation Product Family of engines. After previously working in the power industry, she moved to aerospace, gaining diverse experience at Pratt & Whitney, International Aero Engines and Hamilton Sundstrand. Fraga's contributions continue to positively impact Pratt & Whitney's service programs for legacy and next-generation engines.

LYNN FRAGA
2013

RENEE SUTHERLAND
2013

Renee Sutherland

Renee Sutherland has been a Pratt & Whitney design engineer for the Compression Systems Module Center. Cross-trained as a structural analyst, she also provided support to integrally bladed rotors on the manufacturing floor. Sutherland serves as chairperson to the Southern New England Association of Technical Professionals, a professional chapter of the National Society of Black Engineers (an organization she's worked with for almost 10 years), helping achieve its mission to "increase the number of culturally black responsible engineers who excel academically, succeed professionally and positively impact the community."

Kerry Kozaczuk

Kerry Kozaczuk supports Pratt & Whitney's Materials Engineering lab and, in her spare time, donates her artwork to nonprofit groups supporting equitable practices and respectful treatment of people and animals. Kozaczuk is a member of several diversity initiatives at United Technologies Corporation, and served as chairperson of the Connecticut Stonewall Speakers. She earned a bachelor's degree of fine arts in illustration from the University of Connecticut and an MBA from Rensselaer Polytechnic Institute.

KERRY KOZACZUK
2013

TEMEKA S. WHITE
2013

Temeka S. White

Temeka S. White led global project management efforts for large resource-based projects, along with the change management process for work transitions. She has served as co-chair of the board of directors for All-Start Gymnastics, as well as for fundraising and obtaining grants, to aid competitive gymnasts ages 6-17 in an environment that emphasizes and promotes safety, stresses fundamental preparation, mental and emotional enhancement, competitiveness and good sportsmanship. White has a bachelor's degree in economics and an MBA from Albertus Magnus College.

Renee Jurek

Renee Jurek has been an aerodynamic engineer at Pratt & Whitney since 2005. In her role as "Hot Section" Engineering Durability Computational Fluid Dynamics lead, she was innovative and found technical solutions to complex problems relating to gaspath aerothermal modeling, while keeping the team focused and reporting progress to management and Senior Fellows. She leads by example, delegates when appropriate and negotiates where needed to advance her projects. Jurek is a gifted aerodynamicist as well as being dedicated to various nonprofit activities, mentoring a FIRST Robotics team, the FIRST Lego League and the Power to Read Program.

PENNY CLOFT
2014

RENEE JUREK
2014

Penny Cloft

Penny Cloft led a Lean Product Development initiative as a Senior Fellow Discipline Lead for Systems Engineering at Pratt & Whitney. She is an advocate for improving engineering processes and ensuring they add value to the business and, ultimately, to customer satisfaction. Cloft introduced tools and methods that improve knowledge capture and reuse across programs to ensure system architecture decisions are data-based and include cross-organizational input.

Jan Lin

Jan Lin has been an Integrated Product Team Leader at Pratt & Whitney since January 2012, with numerous contributions to her credit. Previously, she was a design engineer at Cobham in Orchard Park, New York, where she worked on components for the International Space Station. Lin received her bachelor's and master's degrees in mechanical engineering from the State University of New York at Buffalo and holds an MBA from St. Bonaventure University.

JAN LIN
2014

AMY COMER
2014

Amy Comer

Amy Comer, as a manager in Mechanical Systems and Externals, Advanced 6th Generation Fighter Jet Development, participated in a range of product design and development activities including Pratt & Whitney's industry-changing Geared Turbofan™ Development Program and the advanced U.S. fighter program. Comer, a graduate of Cornell University, joined United Technologies in 2002 and has held positions of increasing responsibility at both United Technologies Aerospace Systems and Pratt & Whitney.

ELIZABETH MITCHELL
2014

Shelton Duelm

Shelton Duelm was the design discipline manager for Advanced Combustors and Power Systems at Pratt & Whitney, responsible for advancing combustor design and overseeing approximately 15 design engineers. Her achievements helped position her as a principle engineer for Pratt & Whitney's Next Generation Product Family systems design. Duelm holds a bachelor's degree in mechanical engineering from Worcester Polytechnic Institute and a master's degree in product development engineering from the University of Southern California.

Elizabeth Mitchell

Elizabeth Mitchell has worked on military exhaust programs and co-led export regulation training efforts for her department. She is a 14-year veteran of Pratt & Whitney with experience in controls, management and strategic planning. Her past volunteer work includes founding the Hartford Professionals Chapter of Engineers Without Borders and associated corporate fundraising efforts.

SHELTON DUELM
2014

Lisa Starkey

Lisa Starkey, a 16-year Pratt & Whitney employee, is a Global Services engineering repair manager responsible for leading aftermarket repair strategy. She has advanced her career through assignments across both military and commercial business segments. Starkey holds a bachelor's degree in engineering mechanics from the University of Cincinnati and a master's degree in management from the University of Maryland.

SHARI BUGAJ
2015

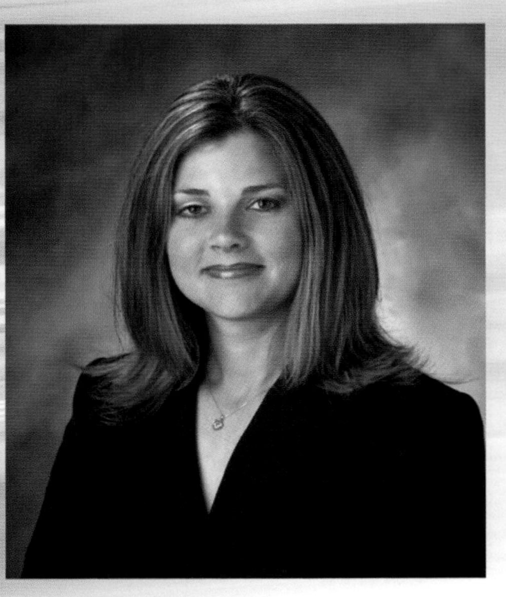

LISA STARKEY
2014

Shari Bugaj

Shari Bugaj, an Engineering Part Family manager, has led the cost-reduction efforts for the Next Generation Product Family through thermoplastics. She also engaged cost-cutting at the company's suppliers by creating workshops to focus on learning best practices in other industries for these materials, as well as developing in-house training modules. As part of her nonprofit work, Bugaj helps plan activities and raise funds for FoodShare, Special Olympics and Interval House.

Anna Patrizzi

Anna Patrizzi demonstrated her leadership skills by inspiring a team with a new vision. In 2012, Pratt & Whitney concluded the largest and most complex acquisition in its history, purchasing Rolls-Royce's share of International Aero Engines. Anna was responsible for ensuring engine part deliveries to support all new engine assembly and significant spare part sales volumes to support more than 5,000 in-service engines. Patrizzi then expanded her leadership prowess and career development as director of Supply Chain Integration, supporting the company's Next Generation Product Family future.

ANNA PATRIZZI
2015

LINDSAY LANDRY
2015

Lindsay Landry

Lindsay Landry, manager of Engineering Integrity Stats & Safety, is an automation and optimization expert, as part of Pratt & Whitney's evolving design systems. She is pioneering the application of these techniques, many for the first time, to real-world product development challenges. Landry has earned a doctorate degree and has years of industry experience combined with energy, motivation and conviction, making her a company innovator.

Carolyn Begnoche

Carolyn Begnoche, a senior designer/drafter, has taken community service to the next level. Her goal? To empower young women with knowledge and through hands-on activities. She is outreach chair for the Society of Women Engineers (SWE), Hartford Section, where she teaches K-12 STEM activities; is an active member of Women In Aviation International; and is a FIRST Robotics judge for New England and a Connecticut Invention Convention judge. Begnoche is pursuing a manufacturing management degree at Central Connecticut State University, where she was elected a SWE counselor.

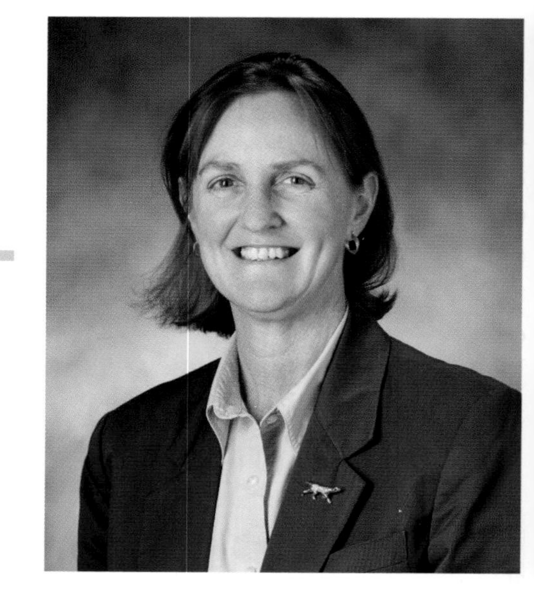

Margaret Steinbugler 2016

Carolyn Begnoche 2016

Margaret Steinbugler

Margaret Steinbugler, a manager, Materials and Structures, has led innovative teams to deliver new technologies and solutions at four United Technologies' divisions. At UTC Power, she led a team that developed a zero-emission fuel cell for transit buses which set world records for fuel cell durability and she is an inventor on nine fuel cell system design patents. At Pratt & Whitney, she led the Materials Engineering team that manufactured the company's first-ever Compression Module Center turbine blades and, in 2015, was recognized on three Leadership Award finalist teams, including one winning team.

Anita Tracy

Anita Tracy, a principal engineer, works on turbine airfoil structures, where she established – and continuously improves – prescriptive methods for conducting structural analysis of gas turbine engine turbine blades. Tracy holds bachelor's, master's and doctorate degrees in aerospace engineering from the University of Maryland. Away from work, she is an alumnae adviser to her sorority chapter at Yale University.

STACY LANATA
2016

ANITA TRACY
2016

Stacy Lanata

Stacy Lanata, a Business Development manager in Aftermarket Engine Services, is tasked with identifying and developing new service offerings to grow the existing business portfolio. She joined Pratt & Whitney in 2000, holding various positions of increasing responsibilities, including her former role as the manager of Turbine Airfoils Strategic Sourcing. She has sought out roles that challenge her skills and take her out of her comfort zone so that she is continuously learning.

Nancy Cika

Nancy Cika is the PW4000 Program manager, responsible for the financial and technical health of a fleet of more than 2,500 engines and 100 global customers. Previously, she was lead negotiator on a number of mergers and acquisitions, and is a Six-Sigma Black Belt and lean practitioner who put those skills to good use when serving in Manufacturing. A leader in youth ministry at her church, Cika aspires to making a difference, learning, challenging the status quo, delivering results, enjoying everything you do, and pushing women and girls to aim high and break the mold.

Nancy Cika
2016

Mary Schubert
2016

Mary Schubert

Mary Schubert, as program manager for the company's United Technologies Product Part Approval Process, is driving incorporation of this strategic initiative to improve quality for key engine programs and developed and related proficiencies to help mainstream a critical methodology across Pratt & Whitney and its supply chain. The role builds on her background in Operations, Quality, Manufacturing Engineering and Aftermarket Engineering. Schubert is driven by a passion to learn and use her technical knowledge to improve the company.

STEP Ahead

The Manufacturing Institute launched the STEP (Science, Technology, Engineering and Production) Ahead – Women in Manufacturing initiative in 2012 to celebrate women in manufacturing who are making a difference through advocacy, engagement, promotion and leadership.

Nationally recognized for leading research efforts to tackle a serious skills gap in manufacturing, especially the underrepresentation of women in the industry, the Institute began promoting the leadership and career opportunities for women in manufacturing.

As an early adopter, since 2014, Pratt & Whitney has recognized numerous women in key manufacturing roles. The company's STEP Ahead - Women in Manufacturing honorees have been representative of the growing STEP talent in its factories and help define the industry's future on a national stage.
STEP Ahead Award honorees have accomplished success within their companies and are proven leaders in the industries in which they operate.

Employee working titles in this section reflect the award recipient's position at the time of their award nomination.

Jonna Gerken
2014

Jonna Gerken

Jonna Gerken has been at the forefront of product development, working with multi-disciplinary teams to create and test prototypes for technology instrumental in next-generation product designs. Her leadership in the Society of Women Engineers has honed her professional skills and a belief that it's essential to champion women in engineering. "I take great pride in being a role model and mentor to new engineers, helping break down barriers and creating a more inclusive work environment," Gerken said.

Meggan Harris

Meggan Harris's passion for manufacturing shines through her exceptional collaboration with suppliers, the mentoring of colleagues and her volunteerism as a founding member of Engineers Without Borders New London County. Her dedication to concurrent engineering, collaborative engineering, process improvement and manufacturing capabilities are unparalleled. "The ability of our nation to produce quality products affordably is what will keep us competitive in the global marketplace," Harris said. "Through continued innovation, we not only develop new jobs, but also give younger generations the opportunity to further technology."

Meggan Harris
2014

MONICA ARIAS
2014

Ruthanne Szumski

Ruthanne Szumski promotes excellence through the development and execution of innovative machining processes delivering critical components to Pratt & Whitney's revolutionary Next-Generation Product Family. Her technical skills and leadership have provided exemplary results including the incorporation of safeguards for developing innovative manufacturing processes. "To have a successful career in manufacturing you need to listen first, then say, then do what you said, and keep listening," Szumski said. "Execution of a new concept is never perfect. Success is in the ability to correct and engage others in the endeavor." Szumski has been recognized as a distinguished alumna by the University of Rochester for her contributions to student career development.

Monica Arias

Monica Arias' technical expertise, project leadership skills and knowledge of lean manufacturing help her to reduce the cost of poor quality and resolve issues through the elimination of process variation. "I'm passionate about manufacturing because it's such an exciting field," Arias said. "It requires technical knowledge while allowing you to be 'hands-on' with your job and doesn't keep you at your desk all day. You get to see things come to life." Since 2005, Arias has shared her skills in the community as an active member of e-Buddies, promoting social inclusion online for people with intellectual and developmental disabilities.

RUTHANNE SZUMSKI
2014

ELLEN McISAAC
2015

Ellen McIsaac

Ellen McIsaac, a composite structures engineer, has a unique background as both a materials and a mechanical engineer, enabling her to examine manufacturing issues from multiple perspectives. She is adept with a range of materials, geometries, operating conditions and manufacturing techniques, allowing her to demonstrate innovative thinking and applications. McIsaac is actively involved in both the Society of Women Engineers, as well as in FIRST Robotics, in which she was recognized in 2014 as Volunteer of the Year for New England. McIsaac is inspired by Lorna Gibson, one of her MIT professors, noting that "She has balanced incredible success in her work across several disciplines of engineering while serving as a leader in the community."

AMANDA VARRICCHIO
2016

Susan Walsh

Susan Walsh's understanding of technical certification and Federal Aviation Administration criteria is vital to the advancement of Pratt & Whitney's Commercial Engine programs. Walsh also has been integrally involved with the Commerce Department, promoting international advocacy for Pratt & Whitney products and driving policy initiatives. Walsh is a board member of the International Aviation Women's Association, the Aero Club of Washington and the Royal Aeronautical Society/Washington, D.C., Branch. "A robust and innovative manufacturing sector is essential to sustaining and expanding domestic economic growth, and delivering wider and increased prosperity to the country," Walsh said.

Amanda Varricchio

Amanda Varricchio has demonstrated outstanding leadership and relentless drive, making her the "go-to" test engineer for Pratt & Whitney's Next Generation Product Family-30k flight test. Varricchio is a mentor to junior test engineers, new hires and interns and fosters an inclusive work environment. As a Daniel Webster College alumna, she has been critical to the company's successful recruiting efforts there. She also shares her talent in support of the Mercy High School Robotics Team. "I am passionate about manufacturing because it is a gateway to innovation," she said. "One simple idea can spark real change. There is no better feeling than when your team meets that key deadline, completes that big test or solves that problem that has been driving you crazy."

SUSAN WALSH
2016

50 Years of Contribution and Counting...

My Story of 50+ Years of Service: Maureen LeClair

Maureen LeClair in 1962 and today with the longest service record of any female employee.

My first day at Pratt & Whitney Aircraft was July 12, 1962. I had graduated from East Hartford High School only three weeks earlier. I started out as a clerk-typist, with a manual typewriter, in Manufacturing Engineering's Plant Layout group in the Podunk plant in South Windsor. I felt like I was learning a whole new vocabulary with words like templates, layouts, and base sheets.

Our group moved to the East Hartford plant in February 1971. In 1976, I was asked if I would like to move into the drafting group—still within Plant Layout. I decided to take the chance, although I knew nothing about drafting. It was on-the-job training. I was taught everything I would need to do my new job. At first, I felt strange sitting at a drafting board instead of a desk. I learned how to draw templates of machines, taking a drawing of a machine that came in from a manufacturer and scaling it down to ¼ inch = 12 inch so the layout men could use them to create layouts of the shop. This information was kept on 3 × 5 inch cards.

The templates and layouts were created on velum with ink. All the lettering was done with a Leroy set. For those who don't know, that is a method of lettering on the drawings. You traced the letters from a template with a scribe onto the velum.

In 1979, we got our first computers and the Computer Vision drafting program. Since that time, we have moved on to AutoCAD. My little note cards made their way into the Lotus program and are now in an Access database.

In the 1990s, Plant Layout was disbanded and I moved into Plant Engineering, taking all my work with me. I tell everyone I am the historian, the librarian. After all these years, I know quite a bit about all these buildings and how they have changed with our products (historian). I am in charge of our archival system, Documentum (librarian).

I really enjoy what I do here at Pratt & Whitney, and I'm glad I've had the opportunity to spend all these years here. Thank you, Pratt & Whitney.

Among the dedicated employees of Pratt & Whitney, these six women stand out for their longevity and dedication. They are the few who have witnessed the progression of women's evolving roles at Pratt & Whitney.

- Bonnie L. Gray
- Martha F. Toler
- Dorothy G. Langly
- Shirley J. Edmonds
- Nancy G. Woods
- Joan U. Olson

It's Only the Beginning

As technology advances, and the printed page becomes less relevant as an archival tool, we will continue to transition to digitally documented storylines about women's professional development.

We will also continue to document the stories of Pratt & Whitney's women employees, whose outstanding achievements can be found across a variety of disciplines. The narrative will go on as Pratt & Whitney approaches its 100-year anniversary and forges into its bright future . . .